高屋建瓴艺海扬帆

丙戌金秋 萧乃素

贺华艺公司成立二十周年

序言

"从历史的观点看,建筑学一直是由实践而非理论所定义的。"在华艺设计成立的二十年中,中国建筑"飞速追赶着西方现代建筑走过的百年历程……在经过足够多的怀疑与盲从,学习与拒绝,遗留下足够多的创造与抄袭,喝彩与批判,也积攒了足够多的坚持与放弃,失落与成功之后,"一切正如Herman Hertzberger所说"建筑学很少能成功地逃避它显然难以逃避的命运——寻求顺应一种或另一种潮流,而不是抛弃表面的时尚与繁华并代之以真正的现实。"

然而,"今天有谁能声言建筑学发展得不好?看起来今天已没有更多的限制……在改变严格约束的召唤下,一切都看似没有限制并处于一种轻率的状态下……现今的建筑学的世界类似于一场足球赛,仅有一些无所不能的球员却没有了球门柱……虽然有华丽的行为,但比赛的趋势和我们真正的期盼却不甚清晰,"到处充斥着似是而非的"包装盒礼物"式的设计。正如加塞特在《大众的反叛》中所说"今天的世界依然缺乏根基,漂泊不定。"

但有一个共同点是毋庸置疑的,建筑师按照人类的期望而进行设计,当然,进一步的讨论会产生诸多的细节:建筑是一门职业还是一门学科?是学术还是艺术?是工艺还是科学?是环境管理还是理论领域,亦或就是一种公司文化……对于业外人士,似乎怎么定义都有可能。有一点很清楚,一个建筑设计机构所做的决定,都不仅仅决定他们完成的作品,还决定其最终成功的可能性。

华艺设计致力于实现"工艺与科学"和"职业与管理"相平衡的"平台化"战略。对比当下对公共注意力和新奇的疯狂追求,华艺更加倾向于头脑清醒和有公众责任感的设计姿态。

公众责任感是华艺企业文化的坚实基础。在华艺,设计、策略、立场是三位一体的。"设计必须从属于设计者介入现实的整体策略,而策略又决定于设计人的立场,这个立场包括政治、经济、文化和社会的立场,脱离开立场和策略的所谓设计是无本之木与无源之水。"当前中国快速的城市化所引发的机会与条件是史无前例的,也难以移植任何既有的理论框架,这要求建筑师必须负担起解决社会问题的责任,尤其是解决那些由于建筑理论与实践自身缺陷所产生的问题。

基于这种责任感,华艺设计作为一个整体,坚持三个方面的基本要素——团队精神、流程卓越与研究发现,这些也是组成华艺企业文化的最坚实的基础之一。

团队精神与团队合作是华艺工作的特色。建筑学是由一个由广泛的专业学科作为基础的多学科体系驱动的职业,不再仅仅依赖于单个的强有力的领导力量,整个团队都需要分担与承受由领导权带来的部分责任和风险,也以此摆脱"作坊式"的观念和模式。这种内部结构使得协作的团队能够承担真正的大项目。"当你有不同的人在努力工作的时候,你才能得到最好的方案"。依托团体实践的哲学,华艺的项目组相信实践、冒险,并且依赖直觉的设计过程,没有任何单一的设计风格或者意见可以起支配作用,高度无等级和分散制度为更大的职业自由和更快的业主回应提供了更大的可能性。团队精神也创造了一个成长的环境，这其中包括人才的集中，业主与经济发展的推动以及持续的学习。华艺创造了一个可以延续多年的学习的理想场所，年轻人惊讶于巨大的资源库以及项目的挑战性，资深者则要对原有的行业知识进行新的诠释，使之与最新的项目类型和规模相匹配。

依托平台的流程管理是华艺效率的保证。尽管规模较大，华艺仍是一家由商业支持的职业驱动的公司，而不是由职业支撑的商业驱动的公司。作为一家创意驱动的公司，无论这创意来自于科学、规划、构筑还是别的什么方面，无论这些创意来自于团队的任何一方，这些创意在带来对大众及客户最出色的服务的同时，也带来了管理上的巨大挑战。要在市场中取得成功，并没有捷径可走，卓越而细分的流程是维系整个架构的关键与纽带。华艺坚信答案是存在于过程中的，没有流程的保证，多学科体系非但不能调整顺相互之间的差异以适应和繁荣整个职业环境，反而容易在工作上相互冲突。在华艺的流程中，团队整天都忙着把抽象的概念发展成可以实施的具体想法，团队中没有人不是设计师，无论是专攻技术、管理、现场还是合同事宜，所有的人都是从设计的角度去思考，项目支持团队认为自己也是设计师，只是碰巧使用合同、电话、传真和电子邮件来进行设计，而设计团队也持同样的看法。在这里，流程融于管理和设计的每一个细节之中，设计自身作为解决问题的方法，仍不能充分解决当今许多项目的需要，对专业多样性的需求凸显了流程的重要作用。专业化的流程意味着年轻的设计师可以把宝贵的时间专注于专业设计上，而资深人员一方面控制着项目的各个主要方面，一方面传播着宝贵的行业知识，这不仅仅使公司拥有专业人才的高效，也使公司更加具有一个强健的商业运营基础。

创造性的研究与发现是华艺设计方法的核心。华艺设计的团队坚信"研究"是一种优于"创作"的姿态，"发现"比"表现"具备更深的内涵。纯粹的建筑学从来就不曾存在，被建造的领域，还不包括互联的领域，要比我们通常用三维系统确定并命名为建筑学的要深广得多。建筑是活生生的，在超越了形式与风格的表象之后，是由各种多元、开放、矛盾的相互关系为此时此地带来的大量意义，在其中，我们深入"研究"并力图"发现"最具价值的部分并由此产生新的关联与意义。在华艺，美学上的乐观主义是缺乏广泛支持的，反而是"空间的状态……这远比空间的形态重要。"设计更多的是一种描绘愿景的能力而非造型的能力，而营造仅仅是最基本的起点，只有置身于整个"此时此地"的可度量和不可度量的要素中，同时涉及公共与私人空间、活动与组织、流动与可达性、特定场地与仪式、资源的平衡，甚至虚拟空间与时间的分配，才有可能取得"发现"并叠合成高质量的愿景，从而超越表象化的风格与形式。华艺方法论认为在寻求设计愿景的过程中，对设计者以及被其设计的建筑的适度抑制是必要的，正如巴克明斯特·福勒所说"all that except me"，也可说是"自由盛行之处就没有决策的可能"。任何个人的立足点都是犹如维特根斯坦所说的关于覆盖于世界的复杂表象之上的有规划的网络系统，是一种自明的内在系统，在某种意义上对外部世界和别人是一种暴政系统。这就是我们强调"发现"而放弃"表现"的缘由，如果过于崇尚自身内在的力量，在设计的整体架构上就会存在缺陷，导致形式上过度的设计和深层技术能力的缺失，前者正如摩西·塞夫迪所言"唯美得丑"，后者则因技术滥用而导致设计项目自身的失控，引发肤浅的建筑表现泛滥而"钟情于细枝末节的自我表现。"

华艺作为创意驱动的公司，对"美"的追求是毋庸置疑的，但是我们坚信"美产生于对别的事物的关注"。美若是故意为之，则易于流于做作、感伤和趣味低下。正如俄罗斯作家约瑟夫·希洛得斯基在《水印》中批评埃兹拉·庞德对于美过于直接的依附，"……也没有打动我，主要的还是老问题，在美后面追随……美不能作为目标，它总是别的什么的副产品，常常是非常平常的探寻。"

在这本作品集中，很难用一种风格或模式为之打上标签，因为我们在华艺所做的，正是"非常平常的探寻"。

华艺设计顾问有限公司
2006年10月

Preface

"Viewed from history, architectonics is always defined by practice but not theory" during the twenty years since Huayi Design has been established, Chinese architecture "swiftly pursue after hundred years process of western modern architecture, lose and success with much more creation and plagiarizing, claim and animadverting and also with enough persisting and abandoning." As Herman Hertzberg said "architectonics scarcely to avoid the fortune difficult to evade successfully to conform one or another trend instead of abandoning exterior fashion and prosper even replace with absolutely reality."

Moreover, "Who can claim architectonics develops weakly today? It seems that there is no more limitation ⋯ Under recalling to alter strict inhabitation, all appears to be located in impetuous status without limitation, nowadays architectonics atmosphere like a football game only with incapable player but no goal ⋯it is not distinct with our expectation and game trends by the gallant behavior", everywhere is full of specious "packaging gift" design. Gasset said in Rebellion "The world today still squander without foundation."

The common ground is evident that architect design upon human expectation, surely the further discussion will arise many details: is architect a professional or a subject? Learning or art? Technology or science? Environment management or theory field? Maybe is a corporation culture, it is possible to define whatever for man in other professionals but one point is clear that the decision made by a architecture mechanism not only decide their works completed but also to the possibility for success ultimately.

Huayi design always applies ourself to realize platform management balanced between "technology and science" and "occupation and management" Compared with insane aspiration to public attention and novelty, Huayi is apt to design attitude with screwed idea and public responsibility.

Public responsibility is solid foundation for Huayi business culture. Design, strategy and standpoint are integrated together in Huayi. "Designer must be subject to intact strategy interposing reality, which decide designer standpoint which include to politics, economy, culture and society, so-called design broken away with standpoint and strategy is water without a source, and a tree without roots" Opportunities and condition arisen from present rapid metropolis building is unprecedented which is difficult to transplant existing theory frame and it need architect to shoulder the responsibility to solve social problems, especially arisen from self defection in architecture theory and practice.

Based on this responsibility, three elements in Huayi design-team spirit, process excellence and research discovery as the integrity to form one of solid foundation for Huayi business culture.

Feature is team spirit combined with team cooperation in Huayi. Architectonics is an occupation extensive professional subjects driven by multi-subjects that no longer depend on single leader might, the entire team need to shoulder and endure part of responsibility and risk brought by leader right to get rid of "workshop" idea and model. This kind of interior structure can make concurrent team to shoulder really large project. "Only do various people strive for qualified work, you will achieve the best project." Upon team practice philosophy, Huayi project team accepts practice, risk, highly classless and distribution scheme provide more possibility for occupation freedom and fast owner response. Team spirit creates a growing atmosphere which includes human resource concentration, drive for owner and economy development as well as consistently learning. Huayi create ideal space to learn lasted for many years, youth is surprised with giant resource library and project challenge, experienced staff must to annotate origin industry knowledge to match with the latest project and scale.

Process management upon platform is guarantee for Huayi efficiency. Although with large scale, Huayi is still a company driven by occupation supported by business instead of driven by business supported by occupation. As inspiration driven entity, no matter it is originated from science, project, building or others or even from any part of the team which bring about the giant challenge to management at the same time to provide distinguished service for public and customers. It has no convenient path to succeed in market, excellent and detailed process is key and ligament. We believe the answer is existed in the process without insurance to process, multi-subject scheme can't adjust difference with one another to adept and promote entire occupational environment, it is easier to conflict in work. In our process, team is always busy to transfer abstract conception into practical idea, no one in the team is designer, no matter is focus on technology, management, site or contract, we all think from design, project support team think they designer to design contract, call, fax and e-mail accidentally, design team with the same idea. Process is integrated with each detail in management and design, to design ourselves still cannot to solve present projects, the need to profession diversity identify important role for process. Professional process means young designer to focus precious time on design, and experienced staff to control various parts of project and at the same time to spread industry knowledge which not only upgrade efficiency for professionals but also to build solid business operation foundation for company.

The core of our design method is creative research and discovery. Huayi team insists on an attitude "research" is superior to "creation", "discovery" possesses more deeply content than "representation". Pure architectonics never exists, the constructed field not including Inter-net field is more extensive than architectonics we confine with three-dimension and named. Architecture is alive is large amount meaning brought forth by various plural, opening, contradiction at this time and space, hereinto we "research" and strive to "discover" the most valuable part and arise new correlation and meaning. In Huayi, optimism to aesthetics is short of extensive support, "space status … is more important than space configuration." Design mostly is capability to describe vision instead of stereotype, and creation is only the basic beginning. It is possible to attain "discovery" and combined with high quality vision to surpass exterior style and format only to indulge into measurable and immeasurable element in entire " this time and space" and involved with distribution to public and personal space, activation and organization , mobile and attainability, specific place and ceremony, resource balance or vertical time and space.. Huayi mythology thinks it is necessary to constrain medially to designer and his construction, like Buckminster Fuller said: "all that except me", also is "it impossible to decide under freedom prevailing" The standpoint for anybody is like planned network system covered over complicated exterior in the world said by Wittgenstein which is self-evident interior system, in some sense it is tyranny system to outside world and others. It is the reason we emphasize "discovery" and give up "exterior", if we advocate our interior power, the entire architecture will exist defect and result in shortcoming to formally transition and deeply technology capability. The former is like Merces Selfridges said to" attaining ugly only beauty", the latter is uncontrolled to design project by technology lavish to begin with peripheral construction deluge and "focus on self-expression with details."

As inspiration driven company by Huayi, it is indubitable to pursue for "beauty", but we insist on "beauty give birth to attention to other object." If we intend to pursue for beauty it will be easy to fall into affectation, sentiment

HUAYI DESIGNING CONSULTANTS LTD.
October, 2006

目录

综合性建筑

- 2 吉林广电中心
- 8 深圳赛格广场
- 10 深圳市福田区图书馆
- 16 深圳市规划大厦
- 20 深圳安联大厦
- 24 深圳发展银行大厦
- 26 深圳麒麟山庄
- 30 北京中国建筑文化中心
- 34 北京大学深圳研究生分院
- 38 南京朗玛国际广场
- 44 广州白云国际会议中心
- 48 深圳宝安26区旧城改造项目
- 52 绵阳科技产业孵化中心
- 56 深圳创维数字研究中心
- 60 深圳市福田区行政办公楼一期
- 62 深圳福田区政府办公楼二期
- 64 深圳高新区软件大厦
- 68 广州联通新时空广场
- 70 深圳翰宇生物医药园
- 72 深圳喜年中心
- 74 深圳华强广场
- 76 厦门观音山国际商务营运中心
- 78 南海大沥体育文化中心
- 82 青岛海军运动广场
- 88 浙江湖州行政中心
- 90 长春世界雕塑公园
- 92 长春雕塑艺术馆
- 96 长春全民健身中心
- 98 深圳大学基础实验室工程
- 102 济南·综合广场
- 104 西安高新区CBD商务公寓楼
- 108 三亚中油大酒店
- 110 绵阳安县罗浮山温泉山庄
- 111 南京审计学院
- 112 深圳国家工商行政管理总局学院
- 116 南京中国药科大学——江宁校区

居住建筑

- 118 深圳星河国际
- 122 南京天泓山庄
- 126 深圳香域中央
- 130 澳门环宇天下
- 132 长春中海水岸春城——莱茵东郡
- 134 苏州中海·半岛华府
- 140 江门中天国际
- 142 深圳水榭花都三期
- 144 深圳华侨城锦绣花园三期
- 148 利群连云港住宅小区
- 150 成都云岭高尔夫别墅
- 154 深圳中航格澜阳光花园
- 156 合肥安高城市天地
- 160 北京昌平兰亭曲水流觞
- 164 广州中海名都
- 166 南京中海·塞纳丽舍
- 168 东莞西湖春晓
- 172 南京苏源颐和美地（南园）
- 174 南京星雨花都
- 176 南京东方天郡
- 178 苏州埃拉国际·自由水岸
- 180 广州光大花园
- 184 成都花样年·花郡
- 186 南昌金域名都
- 190 大连铁龙·动力院景
- 192 南京麒麟山庄
- 194 重庆华宇·渝州新都
- 198 南宁佳得鑫·水晶城
- 200 深圳中海宝安松岗
- 202 东莞凯达华庭
- 204 武汉东湖·香榭水岸
- 206 南京汇林绿洲二期
- 208 深圳中信红树湾花城北地块
- 210 深圳京基御景华城
- 214 青岛海信慧园二期
- 216 上海新江湾C1地块
- 218 长春威尼斯花园
- 222 北京华美橡树岭
- 224 深圳田园居别墅
- 227 深圳银谷别墅
- 228 上海慧芝湖花园
- 230 阳江核电办公科研后勤基地规划
- 232 浙江湖州仁皇山新区城市设计
- 234 青岛李沧区下王埠村
- 236 厦门洪文居住区
- 237 厦航同安T2006G01地块项目

Contents

Complex Building

- 2   JILIN BROADCASTING & TV CENTER
- 8   SEG PLAZA, SHENZHEN
- 10   FUTIAN DISTRICT LIBRARY, SHENZHEN
- 16   SHENZHEN PLANNING BUILDING
- 20   ANLIAN BUILDING, SHENZHEN
- 24   SHENZHEN DEVELOPMENT BANK BUILDING
- 26   KELIN MOUNTAIN VILLA, SHENZHEN
- 30   BEIJING CHINA ARCHITECTURE CULTURAL CENTER
- 34   BEIJING UNIVERSITY SHENZHEN MASTER CAMPUS
- 38   LANGMA INTERNATIONAL PLAZA, NANJING
- 44   BAIYUN INTERNATIONAL CONFERENCE CENTER, GUANGZHOU
- 48   26th ZONE RENOVATION PROJECT, BAOAN, SHENZHEN
- 52   INDUSTRIAL INCUBATION CENTER, SCIENTIFIC
- 56   SKYWORTH DATA RESEARCH CENTER, SHENZHEN
- 60   SHENZHEN FUTIAN DISTRICT ADMINISTRATIVE OFFICE BUILDING PHASE I
- 62   SHENZHEN FUTIAN DISTRICT ADMINISTRATIVE OFFICE BUILDING PHASE II
- 64   HI-TECH SOFTWARE BUILDING, SHENZHEN
- 68   UNICOM NEW TIMES PLAZA, GUANGZHOU
- 70   HANYU CURATORIAL BIOLOGICAL GARDENS, SHENZHEN
- 72   SHENZHEN XINIAN CENTER
- 74   HUAQIANG PLAZA, SHENZHEN
- 76   KWAN-YIN MOUTAIN INTERNATIONAL COMMERCIAL OPERATIONS CENTER, AMOY
- 78   DALI SPORTS & CULTURE CENTER, NANHAI
- 82   QINGDAO NAVY SPORTS PLAZA
- 88   HUZHOU ADMINISTRATIVE CENTER, ZHEJIANG
- 90   CHANGCHUN WORLD SCULPTURE GARDEN
- 92   SCULPTURE ART GALLERY, CHANGCHUN
- 96   SENIOR CITIZENS' FITNESS CENTER, CHANGCHUN
- 98   THE BASIC LABORATORY OF SHENZHEN UNIVERSITY
- 102   INTEGRATED PLAZA, JINAN
- 104   HI-TECH CBD COMMERCIAL APARTMENT, XI'AN
- 108   SANYA ZHONGYOU GRAND HOTEL
- 110   LUOFU MOUNTAIN HOT SPRING VILLA, MIANYANG AN COUNTY
- 111   AUDITING ACADEMY, NANJING
- 112   STATE INDUSTRIAL & COMMERCIAL ADMINISTRATIVE BUREAU INSTITUTE, SHENZHEN
- 116   SCHOOL OF TRADITIONAL CHINESE MEDICINE, JIANGNING BRANCH, NANJING

Residence Building

- 118   XINGHE INTERNATIONAL GARDEN, SHENZHEN
- 122   TIANHONG VILLA, NANJING
- 126   XIANGYU CENTER GARDEN, SHENZHEN
- 130   HEISHAHUAN MIDDLE STREET BLOCK R+R1, MACAU
- 132   CHINA OVERSEAS RHINE GARDEN, CHANGCHUN
- 134   CHINA OVERSEAS ROYAL PENINSULA, SUZHOU
- 140   ZHONGTIAN INTERNATIONAL GARDEN, JIANGMEN
- 142   SHUIXIEHUADU PHASE III, SHENZHEN
- 144   OVERSEAS CHINESE TOWN JINGXIU GARDEN PHASE III, SHENZHEN
- 148   LIQUN LIANYUNGANG RESIDENTIAL DISTRICT
- 150   YUNLING GOLF VILLA, CHENGDU
- 154   ZHONGHANGGELAN SUNSHINE GARDEN, SHENZHEN
- 156   ANGAO CITY GARDEN, HEFEI
- 160   BEIJING ZHONGGUANCUN TECHNOLOGY ZOO CHANGPING GARDEN MATCHING RESIDENCE B PROJECT-LANGTING~QUSHUILIUSHANG
- 164   ZHONGHAI MINGDU, GUANGZHOU
- 166   CHINA OVERSEAS SENALYSHE GARDEN, NANJING
- 168   WEST LAKE SPRING MORNING, DONGGUAN
- 172   NANJING SUYUAN YIHE MEIDI (SOUTH GARDEN)
- 174   XINGYU HUADU, NANJING
- 176   ORIENT TIANJUN GARDEN, NANJING
- 178   AILA INTERNATIONAL LIBERTY BANK, SUZHOU
- 180   EVERBRIGHT GARDEN, GUANGZHOU
- 184   CHENGDU HUAYANGNIAN · FLOWERY COUNTY
- 186   JINYU MINGDU NANCHANG
- 190   DALIAN TIELONG DONG LI YUAN JING
- 192   NANJING KYLIN VILLAGE
- 194   HUAYU YUZHOU XINDU, CHONGQING
- 198   JIADEXIN PLAZA, NANNING
- 200   CHINA OVERSEAS BAO'AN SONGGANG, SHENZHEN
- 202   KAIDA HUA TING, DONGGUAN
- 204   WUHAN EAST LAKE WATER FRONT
- 206   PHASE II OF NANJING HUILIN OASIS
- 208   CITIC MANGROVE BAY FLOWER CITY NORTH PLOT, SHENZHEN
- 210   SHENZHEN JINGJI YUJING HUACHENG
- 214   HISENSE HUI GARDEN PHASE II, QINGDAO
- 216   SHANGHAI XINJIANG BAY C1 PLOT
- 218   VENICE GARDEN, CHANGCHUN
- 222   HUAMEI OAK RESIDENTIAL DISTRICT, BEIJING
- 224   TIANYUAN VILLA, SHENZHEN
- 227   YINGU VILLA, SHENZHEN
- 228   HUIZHIHU GARDEN, SHANGHAI
- 230   PLANNING FOR YANGJIANG NUCLEAR ENERGY ADMINISTRATION, RESEARCH, AND LOGISTICS BASE
- 232   URBAN DESIGN OF HUZHOU RENHUANGSHAN NEW DISTRICT
- 234   LICANG XIAWANGBU VIALLAGE, QINGDAO
- 236   HONGWENJU RESIDENCES, XIAMEN
- 237   XIAHANG TONG AN T2006G01 PLOT PROJECT

综合性建筑
Complex Building

吉林广电中心
JILIN BROADCASTING & TV CENTER

该项目位于长春市东南方向的净月潭旅游经济开发区内，处于卫星路和东盛大街交叉口的东南角。

吉林广电中心一期工程主要包括演播区、广播节目播出及制作区、制作及播出区、新闻中心、办公区、车库、设备及后勤服务区等。

项目被设想为城市媒体公园中的冰雕，成为公园中的有机构成元素。大尺度的地景给人们带来难以抑制的震撼性。地方冰雕艺术在空间三维和意识形态上放大，这是对人文景观的推崇性操作。

绿化围绕基地展开，并渗透到建筑内部，形成局部空气"过滤层"，在一期和二期之间的湖面营造特色景观区，着力体现四季的更替。

建筑和电影擅长运用蒙太奇手法传播信息，设计试图加以运用，不同时间和空间的场景重叠于湖边广场，变幻动感的特殊效果给人们极强的传媒感受。演播厅外的媒体墙也是主要构成元素。

"城市媒体岛"中引入水景，使基地环境表情更加富于变化，独特的环境为外景拍摄提供了良好的场景。

建筑面积：101800m²
Total Floor Area: 101800m²
设计时间：2003-2004 年
Design Period: 2003-2004

The project is located at Jingyuetan Economic Development Zone in Changchun, at the southeast side of the intersection between Hengxing road and Dongshen street.

The first phase of this project consists mainly of studio, production and broadcasting areas, the news center, an office center, and other support facilities. The project has spectacular eye-level view which is considered both as a sculpture and an organic element of City Media Park. This local ice sculpture is symbolically enlarged in three-dimensions to show respect for the human landscape. A greenbelt is spread around the site and blended into the building, where it also acts as air filter. A lake view, constructed between the first phase and the second phase will represent a unique landscape emphasizing each changing season.

Architecture and film are excellent mediums for montage, and this is applied and exhibited in the building's architecture. Different scenes of time and space are shown overlapping above the square, at the side of the lake, which creates a powerful broadcasting feeling. The attractive waterscape of City Media Island makes the landscape diverse and stimulating, and such unique environment creates a perfect set up for film background.

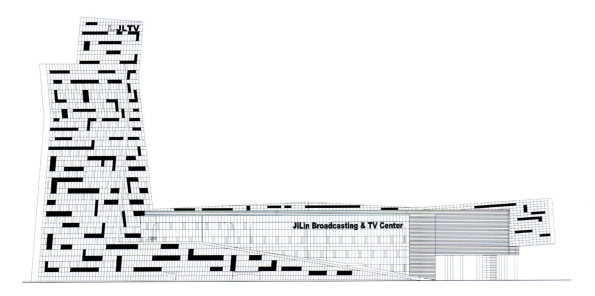

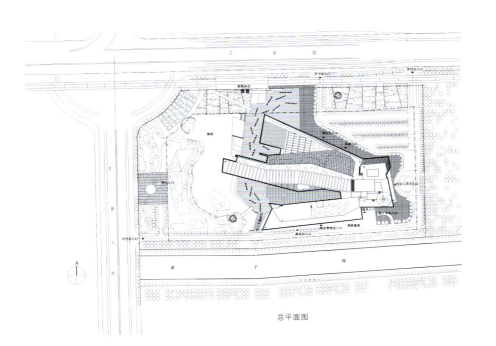

总平面图

一层平面图

深圳赛格广场
SEG PLAZA, SHENZHEN

该项目位于深圳市中心地带、深南中路与华强北路交会处，地理位置优越。

赛格广场主体是现代化多功能智能型写字楼，裙房为10层商业广场，是以电子高科技为主，兼会展、办公、商贸、信息、证券、娱乐为一体的综合性建筑。建筑总高度292.6m，成为深圳市区的重要景观。

塔楼为八边形的平面，外轮廓尺寸为43.2m×43.2m，交通、卫生以及其他附属设施均放置在中心筒内。结构采用了高强度钢管混凝土体系，是当今运用该结构体系的世界最高建筑。塔楼檐口高度292.6m，屋顶天线钢针端高345.8m，在目前世界已建的超高层建筑中高度排名第25。现已成为深圳市区的标志性建筑。外立面主要是灰色玻璃幕墙体，水平方向上装饰有金色外轮廓线条的铝板，体现了简洁富丽的风貌与高雅的气质。塔楼侧面安装有世界上一流的景观电梯，游客可以直达顶部观景大厅，观赏深圳的城市美景。塔楼顶部设置一个圆形直升飞机平台，并高耸起卫星接收天线，这又从另一个角度体现了该建筑的高科技特色。

建筑面积：169459m²
Total Floor Area：169459m²
设计时间：1995－1997年
Design Period：1995－1997
竣工时间：2000年
Completion period：2000

获奖情况 Awards：
2000年度国家科技进步二等奖
2002年深圳市第十届优秀工程设计二等奖
2003年广东省第十一次优秀工程设计二等奖
2003年度建设部部级优秀建筑设计三等奖
2nd prize, "National Award For Technological Advancement", 2000
2nd prize, "Excellent Building Design Award" shenzhen, 2002
2nd prize, "Excellent Building Design Award" Guangdong Province, 2003
3nd prize, "Excellent Building Design Award"; Ministry of Construction, 2003

Located at Shennan Road M. Futian District of Shenzhen city, the 292.6m SEG Plaza is the tallest building in Shenzhen, which becomes the symbol of the city for its complete function of intelligence and informaiton.

The mixed-use development mainly comprises of 10 retail floors for electronic high-tech trading and offices, incorporating with exhibition space, trading floors, information center, stockjobber and entertainment space. The plan takes the form of an octagon with the circulation, sanitation and other supporting facilities located in the central core. Structurally, the building adopts the reinforce concrete structure. The building facade employs the grey curtain wall decorated with horizontal metal cladding which enhances a legible and elegant outlook. The Sightseeing Hall lies on the 71th floor, which is served by high-class glazed lifts, providing dramatic panoramic views of the city.

深圳市福田区图书馆
FUTIAN DISTRICT LIBRARY, SHENZHEN

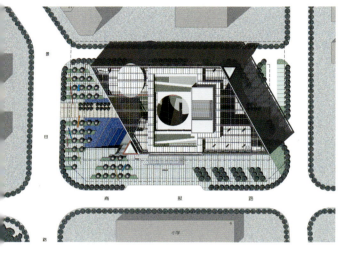

　　该项目地处深圳福田区景田，位于景田路和商报路交会处，北临财政局等高层建筑，南为学校等低层建筑群，西南隔街为狮岭公园。

　　大楼主要由图书馆及信息中心两部分组成，彼此相互独立。两者围绕其中展开，使图书馆具有强烈的向心性，阅览室都面向中庭景观，为读者营造一种宁静优雅的图书阅览及办公氛围。为了丰富中庭空间效果，由西向东做了一系列的退台，空中天桥飞架其中，从城市广场到内部中庭形成一个富于变化的空间序列。

　　本项目对城市街道空间作了周全的考虑。不仅呼应了周边高层建筑的对位关系，还减轻了相互间的压迫感，更重要的是给北面的财政局大楼让出南向空间，并求得一种围合感。由于东、西两侧各让出一三角空间，图书馆自然形成了一个平行四边体，使本方案从商报路及景田路的街景透视都具有强烈的标志性。

　　东、西中庭空间各罩一透空钢构架，在强调建筑轻巧通透的同时，又起到简洁建筑形体、空间界定及遮阳的作用，并在视觉上对城市的喧闹嘈杂起到了隔断作用。光影变化及虚实对比使简洁的建筑形体更加丰富。

建筑面积：48000m²
Total Floor Area：48000m²
设计时间：2002年
Design Period：2002

The site is situated at the junction of Jingtian and Shangbao roads in Jingtian, Futian district, Shenzhen. It is flanked by a high-rise building of the Finance Bureau to the north, a school to the south, and is adjacent to the Shiling Park.

The design program contains a library and an information centre separate from each other. These spaces are arranged around a central atrium providing a quiet and gentle atmosphere for the reading rooms and offices. A distinctive sequence of spatial transformations from the City Piazza to the inner atrium is created with a series of platforms and suspended bridges.

The building respects the primacy of the existing urban space, where harmonious orientation reduces spatial tension between buildings. The plan is a carefully considered parallelogram, that forms two triangular open spaces on the east and west sides. At the same time, it also creates a significant landmark from street perspective.

Steel framed structures on the east and west atriums aims to enhance the legerity and simplicity of the general building shape, as it acts as shading blind and buffer to eschew the bustle of the big city. Moreover, the dynamic shadow cast by the intense Shenzhen sunshine and the distinct contrast produced between spaces also enriches the legible building shape.

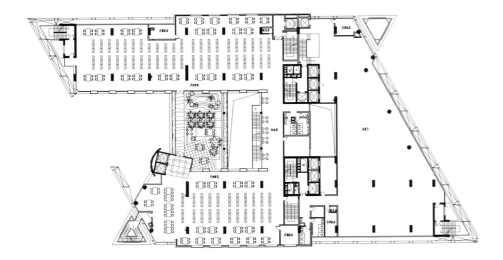

三层平面

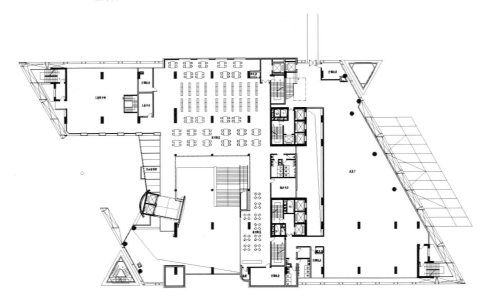

二层平面

深圳市规划大厦
SHENZHEN PLANNING BUILDING

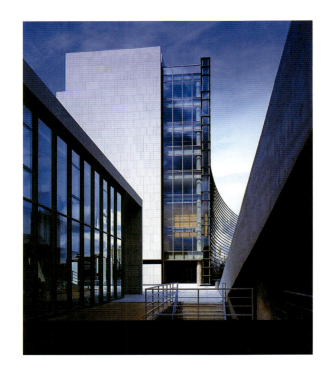

该项目位于深圳市中心区红荔路南侧,用地总面积为13975m²。

规划大厦在使用模式上体现出政府办公建筑是一个秩序化、网络化、程序化以及可持续发展的系统。在形象上体现出政府办公建筑及其机构开放、谦虚、高效、务实与便民的品格。

办公大楼总体造型上力求线条明快、舒展、体块清晰、材质明确。首层窗口办公区域开放、透明,视觉上无障碍,空间流畅、清爽,又便于管理;二层为会议室区域,包括小型会议室6间、贵宾会议室、大型会议室等;三至六层为科室办公区;七层为局长办公区,八层为预留空间及会议室。

建筑面积:34000m²
Total Floor Area:34000m²
设计时间:2001 — 2002 年
Design Period:2001 — 2002
竣工时间:2004 年
Completion Period:2004

合作设计单位:
都市实践
In Cooperation With:
Urbanus

This project site, covering 13975m², is situated at the south side of Hongli road in the central district of Shenzhen.

The Planning Building realizes a well-ordered sequencing application system that is in line with networking requirements and sustainable development goals. It is also transparent and effective, as it is modest and practical. The general building shape is created from clean lines and a combination of clear shapes, with the incorporation of simple, distinct materials. Office areas on the 1st floor eschew visual obstructions in order to keep an open, transparent and fluid business space. The assembly area on the 2nd floor consists of six small-sized meeting rooms, VIP meeting rooms, and large-sized meeting rooms while section offices are located on the 3rd to 6th floors. The 7th floor is where the directors' offices are located and the 8th floor houses the reservation area and conference rooms.

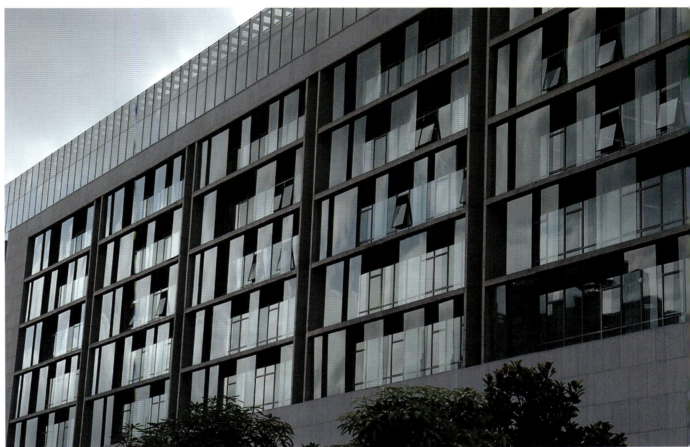

深圳安联大厦
ANLIAN BUILDING, SHENZHEN

　　该项目位于深圳市福田区市民中心东侧，是一幢地下4层，地上35层，建筑主体高150m的超高层写字楼。拥有413个停车位，配有银行、会议室、星级商务中心和其他相应的商务配套设施。

　　该楼采用纯板式结构，在南国的土地上提供了良好的通风条件。整栋大楼设置了风格迥异的28个空中花园，隔层有序地分布在大厦四侧。安联大厦提供了一个会呼吸的花园式办公环境，成为深圳市750万m²中央商务区（CBD）一个重要的商务、办公活动交会点和不可分割的相成部分。

　　建筑师旨在将安联大厦打造成尊重自然、尊重人性、尊重生态的人与自然能够充分交流的活体建筑。

建筑面积：93730m²
Total Floor Area：93730m²
设计时间：2002 — 2004 年
Design Period：2002 — 2004

合作设计单位：
香港王董国际有限公司
In Cooperatin With：
Wong & Tung International LTD.

This project is located at the east side of the Civic Centre in Futian District, Shenzhen. With four floors below ground and thirty-five above, the main body of the building is 150m high. It has 413 parking lots, a bank, meeting rooms, a business quarter, and other support facilities.

For elevation design, the building adopts panel treatments with excellent ventilating attributes. The design aims to create a modern, dynamic building with an effective integration between man and the environment. In the structure itself, 28 aerial gardens are incorporated and logically distributed around the building structure, where a series of green office spaces are created. It also makes the Anlian Tower an integral and organic component for office and business within the CBD district.

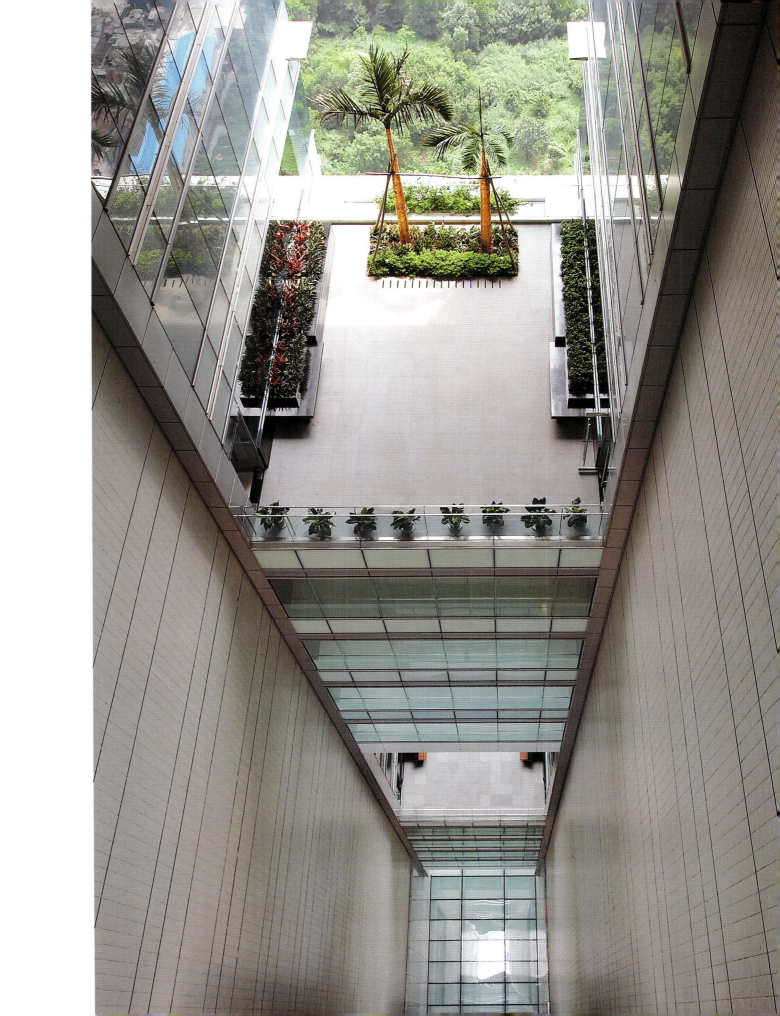

深圳发展银行大厦
SHENZHEN DEVELOPMENT BANK BUILDING

该项目位于深圳市主干道深南大道及蔡屋围金融区之最繁华地段，处于深南路与解放路锐角交会处的独特城市空间中。

基地紧邻金融中心大厦，又是由西向东城市主干道南侧一系列高层建筑的起点。设计以此为契机，将大厦构筑成由西向东步步向上的阶梯体块，辅以倾斜向上的巨大构架，以此寓意"发展向上"，使之成为深圳最具特色的建筑。

设计的风格体现"高技术"的审美趣味，采取超地域的建筑语言，表达一个"当代"的空间形态。设计试图表现改革开放股份制商业银行的独特风格，表现深圳第二个十年的经济发展和发展银行的独特个性。

设计在银行最主要的内部空间中寻求特色，5层高的营业大厅高大宽敞，是深圳所有银行营业大厅中最为壮观的空间。

建筑面积：72334m^2
Total Floor Area: 72334m^2
设计时间：1992 — 1993 年
Design Period: 1992 — 1993
竣工时间：1995 年
Completion Period: 1995

合作设计单位：
澳大利亚柏涛（悉尼）建筑师事务所
In Cooperation With:
Peddle Thorpe Design Office

获奖情况 Awards：
1998 年深圳市第八届优秀工程设计二等奖
1999 年广东省第九次优秀工程设计二等奖
2nd prize in the Eighth "Excellent Design of Building Award" of Shenzhen, 1998
2nd prize in the "Excellent Design of Building Award" of Guangdong, 1999

Sitting in the flourishing region of Shennan Avenue in Caiwuwei financial district, the project site is located in one of the city's most highly defined urban spaces.

The site is adjacent to the Financial Centre Building, where it acts as the starting point for a group of high-rise buildings on the south of the main avenue extending from west to east. In accordance with this special character, the building is defined by setting the body with a ladder shape that ascends steeply to the east as a huge tri-angular frame that makes it a distinctive architecture in Shenzhen city, where it also symbolizes "onward development".

Cutting-edge architecture is presented in its construction through a series of architectural languages from beyond the region to deliver a contemporary space. The aim of this design is to present a unique style for the joint-stock commercial bank and a particular feature that symbolizes economic growth during the second decade of Shenzhen. A business hall five storeys high is located on the ground floor to achieve an atmosphere of spectacular grandeur.

深圳麒麟山庄
KELIN MOUNTAIN VILLA, SHENZHEN

　　该项目是深圳市政府招待所,主要承担接待国家领导人的任务。山庄地处市西北郊麒麟山麓东部的一大片坡地、湖塘,景色非常秀丽。

　　山庄设有综合服务楼、豪华别墅(5栋)、网球场(2个)、游泳池(6个)、沿湖垂钓区、小型游艇码头、山间凉亭、花房等配套设施,形成一个环境优美、高尚、舒适、服务齐全的疗养胜地。山庄的二号和三号别墅在基地南部。别墅依山坡分层跌落,空间上下连通,建筑内外融于大自然之中。别墅内有较大宴会厅、会议厅、休息厅、健身娱乐厅及带有起居室的男女主人房、双人客房、单人客房等,装修高雅,尺度恰当,细部精致。主要房间布置在朝向好、景观好的一侧,有较大开口或平台直通室外专用游泳池。空透的内庭与外景交相辉映,层次丰富。坡屋顶、塔尖塔合使建筑轮廓多姿多态,石砌墙面更具山林淳朴特色。别墅平面与造型结合,功能和地形结合,表现出自然、流畅、清新的属性。

建筑面积:30187m²
Total Floor Area: 30187m²

设计时间:1995 — 1996 年
Design Period: 1995 — 1996

竣工时间:1997 年
Completion Period: 1997

获奖情况 Awards:
1998 年深圳市第八届优秀工程设计二等奖
1998 年深圳市装饰设计作品展三等奖
1999 年广东省第九次优秀工程设计一等奖
2nd prize in the Eighth "Excellent Design of Building Award" of Shenzhen, 1998
3rd price in the "Shenzhen Decoration Design and Production Exhibition", 1998
First prize in the ninth "Excellent Design of Building Award" of Guangdong, 1999

The project is situated in an excellent location, with grand views, on a sloping field east of Kelin Mountain. It is located in the northwest of the Shenzhen where it assumes responsibility for serving visiting dignitaries.

The setting has an integrated tower; luxury villas (5), tennis courts (2), swimming pools (6), view deck, mini dock, greenhouse and other support facilities. It is an open resort offering comprehensive service and amenities in an excellent environmental setting. Two of the villas are situated in the south, clinging to the mountain slope. The villas are elevated on different layouts at different heights, connecting with each other spatially, and penetrating into the nature intimately. The villas with larger sizes and exquisite details are equipped with larger banquet halls, an auditorium, lounge halls, fitness and entertainment halls, and guest rooms with its own amenities. The main hall, which is equipped with a large terrace leading to the swimming pool, enjoys perfect orientation and scenery. The inner courtyard enhances interaction with exterior landscape as it enriches the layouts. Pitched roofing and dormer windows endow the building with an exciting skyline while the stone facade represents simplicity and stability. The amalgamation of plan and shape, function and landform, show the natural, simple and fresh attribute of this project.

北京中国建筑文化中心
BEIJING CHINA ARCHITECTURE CULTURAL CENTER

该项目地处北京西二环与西三环之间的甘家口地区，由建设部及原国家建材局办公楼、会展中心、中国建筑工程总公司办公楼三部分组成。

该项目是集会议、展览、办公、商业于一体的综合性建筑，也是为迎接1999年在北京召开的第20届国际建筑师大会及第21届国际建协代表大会而建。

构思以中国建筑迈向21世纪之"门"为命题，采用了中轴线对称式布局，使中国传统的门、堂、廊等要素得到了合理的体现与发挥；并借鉴门槛的造型，配以壮观的台阶和现代曲面构架大坡度曲线，通过发掘中国传统建筑文化，创作出现代建筑风格。既反映了时代的跨越，又体现了历史的衔接。

建筑面积：82162m²
Total Floor Area：82162m²
设计时间：1995－1996年
Design Period：1995－1996
竣工时间：1999年
Completion period：1999

获奖情况 Awards：
1997年北京市优秀建筑三等奖
2001年度建设部部级优秀建筑设计二等奖
2002年度国家第十届优秀工程设计银质奖
第四届首都建筑设计汇报展建筑艺术创作优秀设计方案一等奖
第四届中国建筑学会优秀建筑结构设计三等奖
3rd prize, "Excellent Building Award", Beijing, 1997
2nd prize, "Excellent Building Design Award", Ministry of Construction, 2001
Silver Medal, "10th Excellent Engineering Design Award", 2002
Excellent scheme design award and First price in the Forth "Report Exhibition of Architectural design of Capital"
3rd price in the forth "China Architecture Association Excellent Architectural Structure Design Award"

The building is situated in Gunjiakou district, Beijing. It holds offices for the Ministry of Construction, State Office for Construction Materials, Head office of China Construction, and an Exhibition centre. Intended to hold the 20th International Architecture Convention and the 21st International Construction Association Convention, the building is a complex and comprehensive construction project combining functions for business and exhibition.

Conceptualized along the theme: 'opening the gates of Chinese architecture for the 21st century', the design layout adopts a symmetrical axis arrangement to demonstrate the logic of Chinese traditional entrance, hall and corridor layout. From a careful analysis of Chinese architecture and layout, a traditional mode of threshold is taken as reference point where familiar Chinese features such as grand entrance steps, curved frames, etc. are incorporated in the design to create a contemporary architecture with a traditional touch. Such a combination not only shows flight over the ages, it also celebrates intimate connection with history.

32 / 33

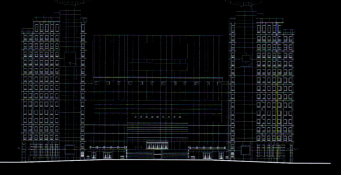

立面图

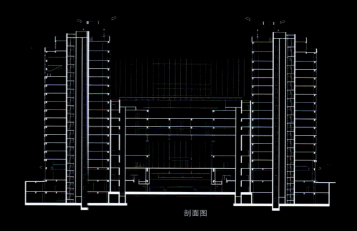

剖面图

北京大学深圳研究生分院
BEIJING UNIVERSITY SHENZHEN MASTER CAMPUS

深圳大学城基地位于深圳市南山区东北部，紧邻野生动物园。北京大学园区位于整个校区的东北部，用地19.88万m²。

项目设计构思沿用了单元式链状集中布局，所有建筑被室外的通廊连接成为一个建筑群，具有相似的建造模式、拼接插接的连接方式；同时，综合考虑室内外空间的多样性与可变性，形成"信息长廊"的设计构思，强调开放共享，汇聚优势，促进各办学实体之间、大学与社会之间的交流。

项目总体上分为办公展览区、学术交流中心、各学科教学区、公共教学区、学生中心与宿舍区6个部分，形成了以中央步行通廊为联系主干，以各个功能区的建筑连廊为骨架，与自然水体、山体地形高度融合的多组团空间结构。同时，在教室间留出空间，将走廊多处放宽，形成集中走道、停留交往平台、开放式中庭等元素于一体的模式，为师生们提供课间交流、活动、休息的场所，也可被充分利用作为各种小型集体活动的空间。这种竖向、水平多处开敞贯通的空间结构也十分适合深圳本地的气候特征。

建筑面积：93000m²
Total Floor Area：93000m²
设计时间：2001—2002 年
Design Period：2001—2002
竣工时间：2003 年
Completion Period：2003

The University Town is located, northeast of Nanshan District in Shenzhen, within the vicinity of Shenzhen Zoo. The project site, named Beijing University Town is located in the northeast of the estate and occupies an area of 198800m².

The layout is comprised of a chain of single buildings linked by an outdoor corridor sharing the same construction and connection technique. Meanwhile, considering the diversity and flexibility of its interior and exterior spaces, the idea of an ´information corridor´ is presented to provide a new public space for academic integration, communication and social exchange between various academic bodies, the university, and society.

The centre is divided into the Office Exhibition Area, Academic Exchange Centre, Specialties Teaching Area, Public Teaching Area, Student Centre and the Student Accommodation Area. Overlooking the whole building group is a spectacular framework forming six functional areas connected by a pedestrian corridor, fittingly incorporated within the natural waterscape and contour. Spaces are integrated between classrooms and corridors are broadened to shape room for passages, platforms and an open atrium. These creative elements serve as leisure places for faculty and students to communicate, relax and hold various social activities. Such vertical and horizontal open spaces and surface treatments also adapts efficiently to the local climate of Shenzhen.

总平面图

南京朗玛国际广场
LANGMA INTERNATIONAL PLAZA, NANJING

该项目位于南京市河西新区CBD与南京奥林匹克公园之间，既是南京市十运会场馆的电视转播主背景之一，也是市政广场和城市绿轴的交会处。

整个CBD的城市设计经过国际竞赛后已有明确的单体建筑设计指引，对建筑的体量、高度、出入口等都有了较严格的规定。

设计强调二元的视觉要素——电视转播中整体视角和市民广场及城市绿轴视角的对立统一。前者要求摒弃一切不必要的细节，展示大尺度、全方位的都市地景，表达以地平线为基准的媒体时代新的宏伟性，后者要求减少视觉与空间的压迫感，形成开敞的绿色视线通廊。同时CBD整体的城市设计不允许出现奇异的建筑体量与轮廓。因此设计抛弃了传统的塔楼、裙房独立设计的模式，而将它们视为水平的和垂直的摩天楼，以线性的体量表达新的宏伟性，并兼顾场所的功能要求和空间定位。

建筑面积：186413m²
Total Floor Area: 186413m²
设计时间：2003 年
Design Period: 2003

This project is located between the CBD of Hexi New District and the Nanjing Olympic Park at the junction of Shizheng Square and the City green belt. It serves as one of the most important backgrounds for the TV broadcast of Nanjing's 10th Sports competitions.

The design framework for single buildings in the city CBD is obtained after an international competition where dimensions and locations of building exits and entrances are stipulated and strictly enforced.

The design for the project emphasizes on the integral visual contrast and harmony between Shizheng Square and City greenbelt during TV broadcast. The former requires reduction of unnecessary detail to show a large-scale, comprehensive, and grand urban space and the latter requires reduction of visual and spatial pressure to form a green alley, conveying a refreshing visual treat for people and the viewers. Moreover, non-conventional buildings are not allowed in the CBD district therefore, traditional towers and annexes are eschewed for a regular skyscraper design showing its grandeur through a simple, linear shape.

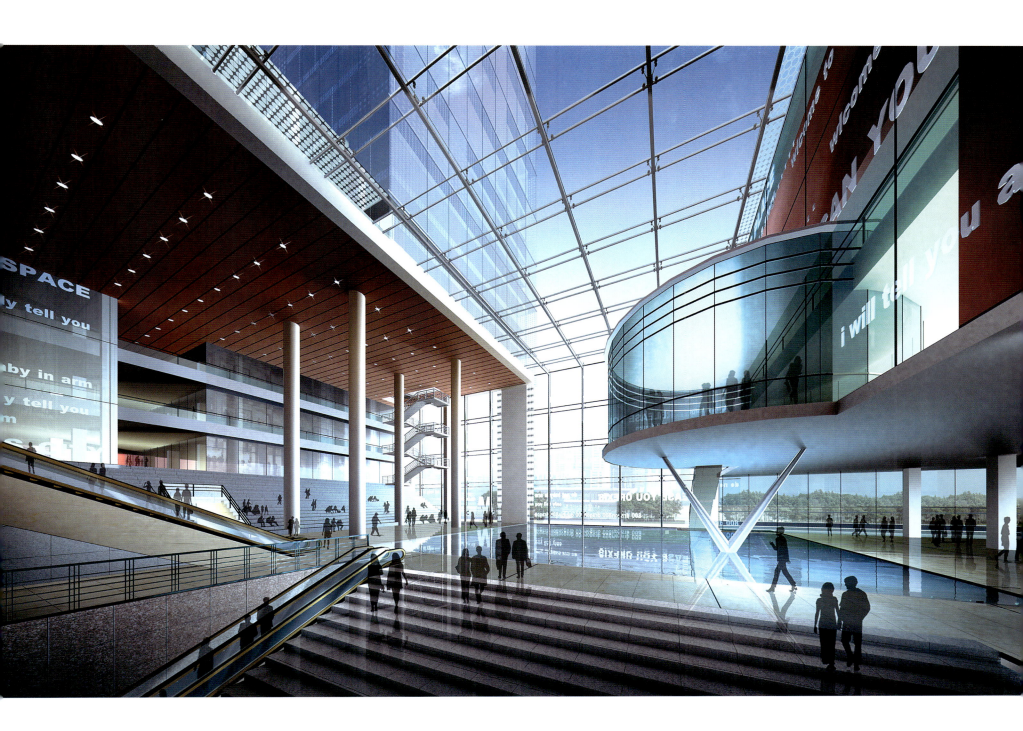

广州白云国际会议中心
BAIYUN INTERNATIONAL CONFERENCE CENTER, GUANGZHOU

该项目的建设选址位于广州市白云山国家风景名胜区西部,未来兴建的白云新城最东端。东南部紧邻云溪生态公园,原址为广州东方乐园,自然环境优美,交通十分便利。基地内外部资源得天独厚。基地面积为 27 万 m²。

项目构思起源于对基地内原有绿化的科学归纳,寻找可建空间,采用岭南传统聚落、院落、廊桥和骑楼的逻辑布局形式,试图营造一个带有浓厚岭南文化特色的内外部空间。建筑形式生态自然,与环境和谐共生,取其寓意为"广合六宇,七星伴月"。

建筑群的整体和局部形成是经过功能原则性组织的结果,13 万 m² 的会议区和 5 万 m² 的酒店区分别占据基地的东南和西北块,既相对独立又保持联系,形成合围之势。同时会议区主体建筑临近鸣泉居出口,便于新老会议区之间交通联系。酒店则靠近新区,便于对外独立经营和管理。

建筑面积:214263m²
Total Floor Area: 214263m²
设计时间:2005 年
Design Period: 2005

The conference centre, which covers an area of 270000m², is sitting at the western part of the Baiyun mountain resort, east of the proposed Baiyun new city. Its southeast border is adjacent to Yunxi Zoology Park, where an excellent environment, convenient traffic access, and splendid natural visual resources are readily available.

The concept of the project comes from a logical conclusion to find space within for the existing natural landscape. Adopting a layout of traditional agglomeration, the design strives to define the interior and exterior spaces with an intensive impression of Cantonese culture. The figure is natural and harmonious with the environment, in line with the adage: "in the centre of the universe, stars accompany the moon".

The form of the whole building group and its components are wrought by a fundamental organizational layout. Large conference areas are located in southeast and northwest of the site, which are simultaneously isolated and connected, forming an enclosed form. The main body of the conference area is near the Mingquanju exit, where it functions as convenient connection between the old and new conference areas. A hotel is located close to the new district for easy operation and management and convenient public access.

深圳宝安 26 区旧城改造项目
26th ZONE RENOVATION PROJECT, BAOAN, SHENZHEN

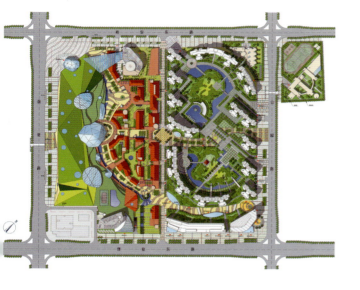

本项目位于宝安26区，紧邻宝安新中心区，四周均有市政道路，交通极其便捷，城市配套设施齐全。在基地东南向，隔路相望是灵芝公园，具有良好的景观资源。整个片区，具备形成商业、酒店、写字楼及居住区的市场条件。

本项目是一个城市综合体，总建筑面积85.8万 m^2，包含住宅、学校、幼儿园、办公楼、酒店、公寓、大型商业、特色商业，其实质是社会产业整合与综合开发的物化体现，目的是为了提升都市生活的安全与舒适度，并提供各种服务及娱乐的都市新型生活方式。

设计合理安排众多功能各异的建筑类型，并处理好之间的相互关系；有效整合地块间的建筑功能，实现两者间的利益最大化；充分利用灵芝公园的景观资源，同时兼顾开发效益，使建筑成为具有标志性的城市综合体形象。

建筑面积：858000m^2
Total floor Area：858000m^2
设计时间：2005 年
Design Period: 2005

The project is located at Zone 26,Baoan District,adjacent to the new city center.It is easily accessible to public transport,social facilities,as well as excellent natural landscape resources.This area has the market condition, ideal for trade,hotel,business and residence.

The 858000m^2 site is an urban complex in Baoan district composed of a residential building, a school, a kindergarten, an office, a hotel, apartments, and commercial units. The goal is conformity and comprehensive development within the society and its industries.A comfortable urban lifestyle with all necessary amenities of serbice and entertainment.

The dedicated masterplan and functional organization combined with the existing resources promote security, a charming living environment, and a sophisticated urban lifestyle that results in a symbolic landmark for the city.

绵阳科技产业孵化中心
INDUSTRIAL INCUBATION CENTER, SCIENTIFIC

该项目位于四川省绵阳市迎宾大道与剑南路西段交叉口东北侧,科技城科教创业园区内。

建筑用地南高北低,较为广阔,与道路高差约1.5m。南临迎宾大道,东接城市广场。基地位置显著,周围景色优美。该建筑形体表达出高科技一体化的纯净。整栋大楼主体结构采用钢筋混凝土结构,外围构架采用钢筋混凝土空间网壳结构,配以格栅,使整个建筑外观统一起来,创造出气势恢宏的效果。是绵阳的标志性建筑。

建筑面积:22541m²
Total Floor Area:22541m²
设计时间:2001年
Design Period:2001
竣工时间:2004年
Completion Period:2004

This project is located at the northeast crossing of Yingbin and west Jiannan roads at the Science and Industry Park of Mianyang, Sichuan province.

The project site has vast land coverage with attractive environment and a fall of 1.5m against the road. It is adjacent to Yingbin road in the south and City Square in the east. In accordance with contemporary high technology, the main body adopts conventional steel and concrete construction while the outer frame adopts a steel and concrete reticulated structure with lattice fencing, aimed at keeping conformity and expressing grandeur. The structure is designed to function as a distinguished architectural landmark for Mianyang City.

深圳创维数字研究中心
SKYWORTH DATA RESEARCH CENTER, SHENZHEN

该项目位于深圳市高新技术园深南大道南侧,紧临30m宽绿化隔离带。

该项目集办公、研究开发及产品展示等功能于一体,体现创维集团屹立于高科技企业之林,勇于开拓进取的企业文化特征。

设计以空间的变化寻求与城市环境的对话。大厦中设置一个巨大的虚空间,模糊外部环境空间与建筑空间的界面,取得两者相互交融,并赋予建筑自身表现力。现代高效的办公环境应是"以人为本"的空间,建筑中设有一个绿色中庭——数码广场中庭,将自然融入办公环境,创造了独特的景观和开放的办公空间模式。

建筑面积:62342m²
Total Floor Area:62342m²
设计时间:2000年
Design Period:2000
竣工时间:2002年
Completion Period:2002

获奖情况 Awards:
2004年深圳市第十一届优秀工程设计二等奖
2004年度全国优秀工程银质奖
2005年广东省第十二次优秀工程设计二等奖
2nd prize,"11th Excellent Building Design Award" shenzhen, 2004
Silver Medal, "National Excellent Engineering Design Award", 2004
2nd prize,12th "Excellent Building Design Award" for Guangdong, 2005

Situated along the south of Shennan road, within the Shenzhen High-tech Park and adjacent to a 30 m-wide greenbelt buffer strip, the project assembles the functions of business, research and development, and product showcase in a unified structure that embodies the distinguished corporate image of the Skyworth Group of companies.

The design of this project seeks to celebrate spatial transformations while integrating with the urban scenery. A big virtual space is located within the building to blend the exterior and interior environment into each other, while giving the structure a unique, identifiable form. The project uses a 'people oriented' design principle, so focus is essentially on providing an excellent working environment that opens up to nature. To create a garden within a building, the Digital Plaza Atrium is established to create unique scenery and introduce a new type of workspace where ecology is part of the office environment.

深圳市福田区行政办公楼一期
SHENZHEN FUTIAN DISTRICT ADMINISTRATIVE OFFICE BUILDING PHASE I

该项目位于深圳市福田中心区西南面、驻港部队基地北面。

它是一座由800个座位会堂、300个座位会议厅、多功能厅及健身、娱乐、办公等用房组成的集会议、休憩、娱乐、办公于一体的多功能建筑。

总体布局吸收中国传统轴线对称手法,延续城市中心区设计思想,突出南北轴线。整个办公综合体由五幢建筑组成,呈梅花形布置,中央为主体,办公建筑起到统率全局的作用,建筑造型庄重大方,既体现政府的崇高权威和凝聚力,又表现出民主、开放的精神和鲜明的时代特色。

建筑面积:68173m²
Total Floor Area: 68173m²
设计时间:1994 — 1996 年
Design Period: 1994 — 1996
竣工时间:1998 年
Completion Period: 1998

The project is located in the southwest of central Futian district, north of the People's Liberation Army Hong Kong Garrison. This government building serves a multitude of functions: an 800 seat auditorium, a 300 seat meeting room, halls, fitting rooms, recreation rooms, and office spaces. It is a one-stop shop government complex engaged in work, business meetings, relaxation, and entertainment.

The entire layout absorbs the symmetrical treatment from conventional Chinese Architecture and it also pursues the design principle of a city centre node to reveal the south and north axes. The complex consists of five buildings arranged in a cinquefoil shape, enclosing a main body at its core and creating a series of perimeter office rooms to command the general space. The design is an attempt to express a stately and grand image which embodies not only the sublime power and cohesion of the government, but the modern spirit of democracy as well.

深圳福田区政府办公楼二期
SHENZHEN FUTIAN DISTRICT ADMINISTRATIVE OFFICE BUILDING PHASE II

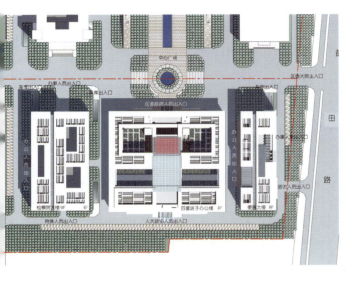

福田区委便民中心项目用地北临福民路，东临益田路及皇岗公园，西临石厦路，项目总用地面积为3.2万m²。

设计注重于生态及节能办公模式的探讨，希望产生一个以亲地办公为主题的生态院落式行政办公建筑，同时吸取中国古建筑的布局精华，采用合院组合式布局。在造型上希望是简洁大气、体量突出而内部通灵。总体建筑群与一期和谐统一的同时，又形成自己完整体系。便民中心的功能属性十分明确，检察院完全独立，会议中心与四大班子办公楼结合设置于中部，突出核心地位，而其内部亦相对独立。检察院和便民中心大楼在功能形象上服务于中部四套班子办公楼，使新建筑的性格表达鲜明。便民窗口办公楼靠近益田路，外来办事人员可直接办事即可离开。各分区位置明确，主次得当，使用便利。

建筑面积：86483m²
Total Floor Area：86483m²
设计时间：2005年
Completion Period：2005

This project is the second phase of the Shenzhen Futian District Administrative Office Building covering an area of 32,000m².

Ecology and energy are the key points considered in this design. To fulfil the demands of an ecological 'gardenesque' office space, the whole layout is designed like a laced courtyard which is a prime virtue of ancient Chinese architecture. An elegant appearance and outstanding building form is achieved from a flexible interior space. The whole building structure forms its own integrated system while being harmonious with Phase I.

The function of this project is evident and demonstrated by a clear organizational layout: The People's Procurement office is technically segregated; the conference centre and office building for four leading branches of the district government are located at the inner core position, while the core itself is relatively unattached. The civil-service office is situated along Yitian road, where it provides convenient access and circulation for the general public.

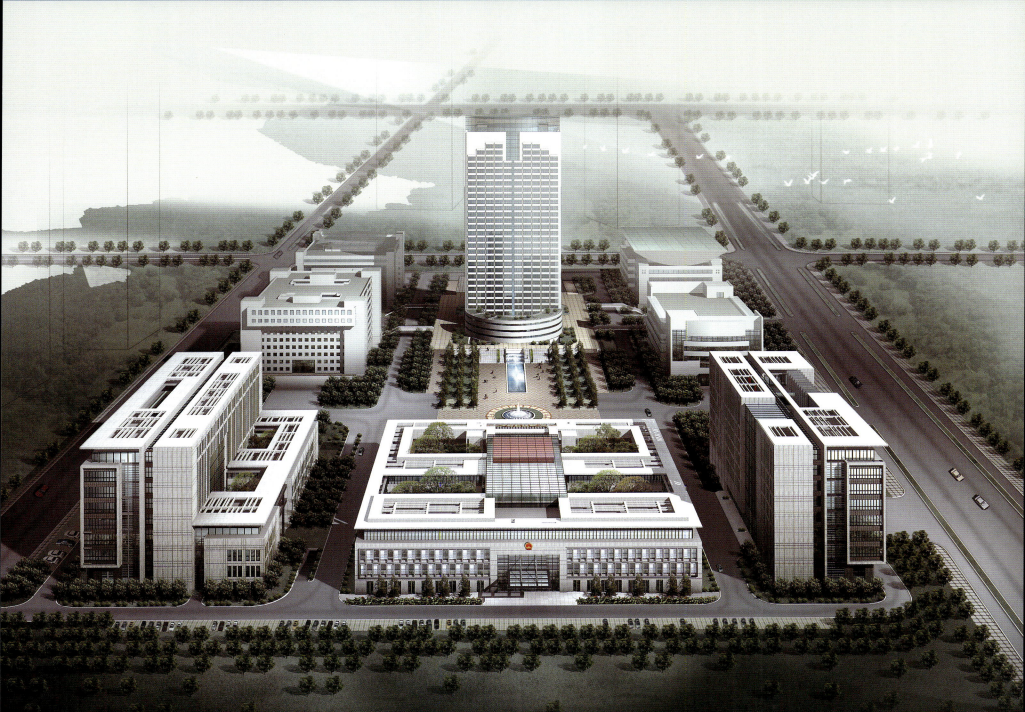

深圳高新区软件大厦
HI-TECH SOFTWARE BUILDING, SHENZHEN

项目用地位于深圳南山高新区中区西片。用地面积26650m²。本项目为国家级重点软件建设工程，建成后将为深圳中小软件企业提供良好的孵化场地，为软件开发和出口提供专业技术支撑和服务平台。

本案最大特征是要求在大厦内设置80个大小一致功能相同的孵化单元，面对相同的元素，致力于从环境、空间和造型上将80个单元建立起一种"计算机程序"的秩序感。采用树形结构和模块化设计理念。80个功能相同的基本单元经过重组和生长形成的多样树形空间体系揭示了数字信息和基因理论盛行时代事物的最简单本质。工作于其中的软件工程师将从富有秩序而又极简的空间、时间组成中获得灵感，萌生新的软件编码。

大楼内包括孵化单元，培训中心，综合数字信息服务平台和后勤用房。

建筑面积：58850m²
Total Floor Area：58850 m²
设计时间：2005 年
Design Period：2005

The project site, with an area of 26650m², is located to the west of the Nanshan High-tech Industrial Park, in Shenzhen. It is the state central software building that functions as a base where small and medium-sized software enterprises can provide professional technology services for the development and export of software. The building contains hatch units, training centres, an integrated digital information service centre, and rooms for logistics.

The outstanding feature of this design is demonstratedd by the location of 80 hatch units with the same functions and sizes. These hatch units are carefully designed in response to the environment, lot space and form. A tree-like planning structure and modular layout design make up the basic framework of this project which strives to reveal a complete essence of the digital information era. The main idea is to provide an orderly and clear working environment to create inspiration for future technological breakthroughs.

广州联通新时空广场
UNICOM NEW TIMES PLAZA, GUANGZHOU

　　该项目位于广州市天河区黄埔大道南侧，毗邻暨南大学、赛马场和珠江新城，地理位置极佳。

　　建筑造型以极少主义为设计原则，运用严谨的逻辑概念、理性的思维方法，突破传统的体形塑造手法，强调"网格"与"窗"的虚实对比，通过网格构架的肌理与变化赋予建筑独特的表情，既有城市界面的延续，又是视觉的焦点。在构架与外墙之间设置可控制亮度和色差的光源，建筑在夜间也能呈现多姿多彩的变化，体现出其卓而不凡的气质。

　　广场设计以反映信息时代特点，以代表信息通信的符号为母题，塑造出一个充满超时空感观的灯光广场。

建筑面积：64308m²
Total Floor Area: 64308m²
设计时间：2003 年
Design Period: 2003

This new plaza is located at the south of Huangpu road in Tianhe district, Guangzhou. It sits on a prominent block adjacent to Jinan University, a racetrack, and the Zhujiang new city centre. The design strives to shape a radiant plaza imbued with a modern atmosphere that reflects the characteristics of the information age.

An adaptation of the classic 'less is more' design theme is carried out logically and rationally throughout the project. It breaches traditional treatments by accentuating a dynamic contrast between the grid structure and the window to reveal a distinctive, stimulating new expression. Furthermore, it extends the transition between the building and the city, while creating a new landmark for the city. The controllable and chromatic illuminations, which are installed between the frame and external walls, enhance the outstanding temperament and diversified feature of the architecture.

深圳翰宇生物医药园
HANYU CURATORIAL BIOLOGICAL GARDENS, SHENZHEN

该项目位于南山高新技术开发区高新中二路与科技中二路交叉口之西北角，基地之西及北面均为将建小区，基地以东和科技中二路一路之隔，为本高新技术区之中心绿化，环境优美。

总体布局将办公部分设于高台，远离路口，居高临下，与制剂车间围合，面对倾斜的大片绿地。整体造型取简洁明快的现代建筑风格，办公科研楼与制剂车间浑然一体。办公建筑与车间的锐角交接处以较大的虚空间过渡，避免体形过长带来的封闭感，造型生动，又具识别性。办公主体墙面以格架饰面，具有强烈的光影变幻效果，立面上的退台与凹廊空中花园创造的绿化虚空间改善了工作环境。外立面富有节奏，又有变幻。一层的南面架空部分和引入的底层绿化，使建筑具有亚热带风情。制剂车间由于工艺和内部的超净需要不作较大的开口面，只在东部的三条转折处做了少量弧面玻璃，与办公建筑的东南弧墙相呼应。大墙面做了部分带形条窗和办公建筑的横带协调。

建筑面积：12200m²
Total Floor Area：12200m²
设计时间：2000－2002年
Design Period：2000－2002
竣工时间：2003年
Completion Period：2003

The site is located at the junction of Gaoxin and Keji 2nd roads in Nanshan High-Tech Development Zone, Shenzhen. To its the west and its north are zoned residential properties and to its the east is a central greenbelt providing environmental amenities for the whole area.

Conceptualised with forthright simplicity, the office for Scientific Research and the laboratory are carefully integrated with each other. The transition between the office and workshop is achieved with careful positioning of a virtual space, adding a degree of richness over the rhythm of the whole building while avoiding a sense of monotony due to the length. The office section facing a sloped greenbelt sits on a platform apart from the road, adjacent to the laboratory. An improved work environment is also obtained through the use of recessed platforms and gardens on the concave corridor. The exterior wall of the office building has latticework finish, which adds to the shading effect and creates an elevation that is both rhythmic and fascinating.

The empty space in the south of the ground floor and the landscape design promises subtropical appeal to the building. Due to technological and sanitary requirements, three corners of the laboratory is covered with a curved glass, instead of a big opening, which also reflects the curve in the south eastern wall of the office building. The massive wall is equipped with partial ribbon windows to coincide with the horizontal moulding of the office building.

深圳喜年中心
SHENZHEN XINIAN CENTER

该项目位于深圳市福田车公庙北片区，处于深南大道和泰然九路交会处的西南侧，地理位置优越。

立面设计为大面积连续落地窗，不仅为办公人员提供充裕的观景采光面，并且是公司展示形象的窗口。通过大面积玻璃窗，使公司的办公空间创意、人员活动、品牌形象等都能呈半公开化，成为一个个直接生动的公司形象展示屏。

该建筑采用节能通风窗代替传统的玻璃幕墙，以实现自然通风，满足室内对新鲜空气的需求，并能防止不良气流进入室内。可自由控制的遮阳格栅能有效解决本地日晒问题，结合大厦本身专门的空调系统设计，既可使大厦实现低成本使用，生态环保，又能保证立面造型美观一致。

建筑面积：51713m²
Total Floor Area：51713m²

设计时间：2000 — 2001 年
Design Period：2000 — 2001

竣工时间：2002 年
Completion Period：2002

获奖情况 Awards：
2004 年深圳市第十一届优秀工程设计三等奖
3rd prize，"11th Excellent Building Design Award"shenzhen，2004

This project is prominently located in the northern area of Chegongmiao, Futian District, in Shenzhen. It has a prominent location at the south western side of the junction between Shennan road and 9th Tairan road.

The building facade is glazed with large-sized curtain walls to provide abundant day lighting and to serve as a ´display window´ exhibiting the corporate image of the business establishment within. The semi-transparent facade visually manifests the creative office space, human activities, and brand images within the building and draws a unique spectacle to the passers-by.

Environmental sustainability and energy efficiency are the key points of the design. A low energy consumption ventilation system is used to enable fresh air circulation and to purge airborne pollutants. The building incorporates an eco-friendly air-conditioning system, as well as adjustable shading blinds to control solar gain and glare. The design achieves the energy conservation requirement as it maintains a spectacular appearance.

深圳华强广场
HUAQIANG PLAZA, SHENZHEN

总体布局：着重关注周边住宅、商业空间联系和主塔标志性的问题：

1. 将商业裙房紧靠主街，并设置广场内街刻意模糊建筑与城市的界限。2. 将四栋住宅塔楼适当远离商业街，并直接排列形成完整的空间序列。3. 主塔直接落地临广场，并使住宅后退，从深南路以及振华路东西端都能直接看到主塔楼的优雅轮廓。

城市设计：注重建筑群体、广场、街道的集合造型，空间上形成南北个性迥异的广场，动线在水平和竖直两个维度展开，平台与大台阶双重结合提供各种活动场所。

可持续发展：1. 观光电梯、空中花园形成有利于节能的基本格局。2. 裙房在双曲面玻璃顶下的公共空间全部自然通风采光，并将阳光引入地下空间。3. 程序控制玻璃顶可以对气候变化作出相应反应，形成街市中的"生物核"，降低能源负荷。

建筑面积：226656m²
Total Floor Area：226656m²
设计时间：1999 年
Design Period：1999

The overall planning emphasizes on the spatial connection of residential and commercial quarters and the symbolic characteristic of the main tower.

The following are key design considerations:
1) Locating the commercial podium on the main street, and inserting an inner street to blur the transition of the architecture and the city. 2) Four residential buildings are arranged consecutively in a sequence of spaces whilst keeping a proper distance away from the hustle and bustle of the main street. 3) The principal tower is set directly beside the piazza where it strikes an elegant posture that is easily identified from both Shennan road and Zhenghua road.

Urban design emphasizes on the aggregate massing of building groups, plazas and archetype street architecture. Spatially, two piazzas shaped in two different styles are placed in the south and north nodes, spread horizontally and vertically. A combination of platforms and structure setbacks assist in providing various activity spaces.

Ecological considerations: 1. Sight-seeing lifts and hanging gardens are incorporated in the basic structure for energy conservation. 2. All public spaces at the podium area are covered and glazed with hyperboloid glass which can supplement natural aeration and optimize light ventilation in the podium and the basement. 3. Programme-controlled glass roofing reacts according to climate variations to decrease the energy burden.

厦门观音山国际商务营运中心
KWAN-YIN MOUTAIN INTERNATIONAL COMMERCIAL OPERATIONS CENTER, AMOY

观音山国际商务营运中心与金门岛遥遥相对，北接厦门机场和厦门本岛通翔安的海底隧道，南邻国际会展中心、国际网球中心与游艇俱乐部，规划用地43.4万m²，总建筑面积138.4万m²，在交通、景观、用地等环节具有先天的优势，并成为厦门市优先发展的核心商务区。

规划利用原有地形营造一个从观音山步行到滨海区的人工生态地景建筑综合体，解决办公楼的配套服务设施和生态景观系统通达的矛盾，突出山海主题。

为使土地利用价值最大化，设计了无障碍通行的架空步行综合系统，以形成飘浮在空中的"第二个地面层"，解决了狭小区域内步行与车行交通的交叉干扰，使其串连上一个个复合的商业、办公与配套服务功能节点，营造出类似漂浮在水面上的船甲板一样的效果。

在功能设置上集商务、餐饮、休闲、娱乐、购物等功能为一体，并且随着使用与市场的变化有许多弹性可变空间，形成了一个弹性线型商业体系，使城市活力与商业价值得以充分体现。

在总体规划上参照了原规划对高层建筑天际线的控制曲线，形成一个船队扬帆远航的建筑形象，生动活泼，从而营造出一个充满都市魅力的现代化CBD城市的繁华景象。

建筑面积：350000m²
Total Floor Area: 350000m²

设计时间：2006年
Design Period: 2006

合作设计单位：
浙江省建筑设计研究院
In Cooperation with:
Zhejiang Architectural Design & Research Institute

Enjoying the natural advantages in traffic, landscape, and location, this scheme will become the core commercial district of the preferential development project for Amoy city. It faces Jinmen Island and shares boundaries with the Amoy Airport and a tunnel in the north. To its south are the International Exhibition Centre, International Tennis Centre and the yacht club.

The design concept is based upon utilizing the existing landform along with the theme of ´mountain and sea´. A man-made ecological complex is created running from the Kwan-yin Mountain to the seaside to facilitate integration of support facilities and connection with the ecological system. To maximize land usage, a second floor layout is created by installing an elevated pedestrian system. The effective pedestrian system resolves the interference of pedestrian and vehicular traffic in such a narrow community. It also enhances the degree of intercommunication within the complex commercial quarter, office areas, and supporting facilities.

The functions that combine business, food, leisure, entertainment and shopping are made adaptable to fit a complex market and eventually, to create a flexible commercial system driven with vitality and value. In accordance with height restrictions, the image of a sailing fleet is realized which results in a lively and flourishing vision of a modern CBD.

南海大沥体育文化中心
DALI SPORTS & CULTURE CENTER, NANHAI

南海体育文化中心项目是一个以体育运动为主的综合建筑群体，功能上集成了体育、图书、展览、商业等。这些功能都具有较强的公众性，因此对其进行一体化设计，以富有流动性的公共空间联系各功能个体，强调整体的融合和资源的共享。同时，建筑与场地作为一个整体来进行考虑，室外空间作为对于建筑室内空间的延展，不仅符合南海地区的自然条件，也烘托了建筑主体，延绵起伏的绿地、平展轻柔的水面使整个建筑群成为环境优美的体育文化公园，使之更加符合新时代全民体育文化发展的趋势。

建筑面积：34184.4m²
Total Floor Area：34184.4m²
设计时间：2005年
Design Period：2005

It is an integrated complex dedicated to athletic sports and on providing support services for sports. It also houses books, an exhibition centre and a commercial centre. Because of its the strong public functions, an integrative design approach is applied to stress the overall amalgamation and sharing of resource by connecting each functional unit through a fluid public space. Moreover, the tact of integrating the architecture and the site as one, and using the exterior space as an extension of the inner space does not only accord with the natural environment of Nanhai, it also complements the principal body. The continuous and fluctuant green land and the quiet water surface decorate the complex as an athletic culture park, which is the current trend in sports and culture development.

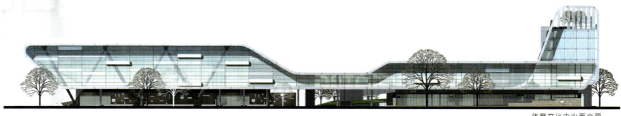

体育文化中心西立面

体育文化中心北立面

青岛海军运动广场
QINGDAO NAVY SPORTS PLAZA

该项目位于青岛市东部沿海，北临城市主干道东海路，东南临大海，海岸线大约656m，有着优良的景观及度假资源。设计力求在规划结构上符合区域规划的要求，优化提升地域环境品质，强化资源共享，为城市提供一个自然、生态、品质优良、功能完备的具有新型滨海城市生活理念的休闲度假区。

规划功能分为酒店、酒店式公寓及海事博物馆三个部分。功能分布结合滨海步行道，分别与海事博物馆及酒店形成两个城市节点，极大地丰富了滨海景观的趣味性、功能性，而酒店式公寓则依靠峭壁沿滨海步行道展开，留出较开阔的沿海空间，有效保护自然海岸线，增加了休闲游览空间，同时可以减少公共区域对酒店式公寓的干扰，而且，公寓建筑依山面海，与步行道可以互为景观。

建筑依山面海留出了较开阔的沿海空间，上面广植植被，设置休闲游览设施，提高建筑环境质量，同时减少公共区域对酒店式公寓的干扰，并且在建筑最顶层设置非上人式的空中花园，加强楼顶绿化，满足城市形象的要求。充分利用环境优势以及原生地貌，使得酒店及酒店公寓每套房间均可以俯瞰大海景观，同时宽阔的沿海绿化休闲空间及滨海步行道也增加了景观的层次。

总建筑面积：46473m²
Total Floor Area：46473m²
设计时间：2005 年
Design Period：2005

The project, with excellent scenery and tourism resources, is situated on the east coast of Qingdao. It sits adjacent to Donghai road in north, and the ocean to the southeast. In accordance with the the master plan, the design aims to improve the existing environment and improve resource sharing. A new version of a coastal vacation area emerges through non-extractive manipulation of natural resource and optimum use of existing ecology.

The overall planning program is divided into the hotel, a hotel-style apartment and a maritime museum. Two city nodes are formed by functional arrangement of networks, thereby enriching interest on and function of the coastal community. The hotel which clings to a mountain face is spread along the coastal pedestrian lane to present an open leisure space and to protect the natural coastline. It also reduces public disturbance to the residential apartments. The air-garden and landscaping on the top floor is designed to cover the requirement for a distinct urban image. Optimum use of environment and existing contour offers each unit opportunity to enjoy delightful views into the sea and to the diversified layouts of landscaping.

浙江湖州行政中心
HUZHOU ADMINISTRATIVE CENTER, ZHEJIANG

该项目位于湖州市仁皇山新区核心区的中心位置,为未来市中心区乃至湖州市迈向21世纪的象征。

设计理念突出以人为本的设计思想,创造具有亲民形象和市民认同感的建筑空间环境。在强调景观性的同时,注重了平易性和公众参与性。从整体环境着手,展现前后广场、内外庭园、绿地、水体的综合效益,体现可进入性和交通便利性。将自然环境引入建筑空间,与江南山水清远之气象相融汇,形成独特的行政中心环境风貌。

总体布局以城市总体规划所确定的飞英塔——弁山法华寺景观主轴为中心轴线,居中为主峰之主楼即行政中心的主要构成——市委、市政府、人大、政协既分且合地一字展开,后勤中心和档案馆分列东西两侧,主楼中心如敞开之"大门"沟通新区旧城,连贯山水景观;众人议事之会议中心置于"门"后中央。南北均临城市干道。

建筑面积:61683m²
Total Floor Area:61683m²
设计时间:2000 — 2001 年
Design Period:2000 — 2001
竣工时间:2002 年
Completion Period:2002

The project is located in the centre of Huzhou Renhuang Mountain District at the Zhejiang Huzhou Administrative Centre. It aims to become a symbolic building to represent the flourishing future of the 21st century Huzhou city centre.

The fundamental design concept is based on civic orientation that endows the space with a sociable image and an identity unique to its constituents. The design considers the whole environment and manifests its comprehensive physiological benefits at the front and rear plazas, interior and exterior gardens, as it improves access to public facilities. Having introduced the natural landscape into the architecture, the space develops harmony with the mountains and the water resource to form a unique environment for the administrative centre.

The planning program takes Feiying Pagoda (Moushan Fahua Temple) as the core axis proposed in the master plan, creating an administrative hub in the centre that serves as main facility for the functions of the Committee of Huzhou, the Municipal Government, the National People's Congress and the CPPCC. The centre is assisted by a logistic hub with archiving spaces at the sides and a conference centre at the rear. The building is conceived as a gateway to connect new cities to the old, and link sceneries of water and mountains to the project site.

长春世界雕塑公园
CHANGCHUN WORLD SCULPTURE GARDEN

作为长春三大艺术园之一的世界雕塑公园,其占地范围、开发规模之大,在国内同类文化设施中首屈一指。其规划设计的主题定位是"未来文化的自然山水园林式现代雕塑公园"。

设计特点:1.规划布局以现代园林式为主。2.规划设计图形模式吸取西方现代文化。3.注重文化性的同时,强调其文化艺术活动的大众参与性,体现以游客活动为中心的设计理念。

设计手法:1.以线型及组团分割空间,在"友谊、和平、春天"的主题之下,划分不同内涵的主题性园区,以强化雕塑公园大布局的目的性、可塑造性和生动性。2.用地大,雕塑小,为雕塑提供多样的展示背景及形式。3.环境设计以绿化为主体,并创造一定的水体景观。4.采用较为强烈的构图形式,突出中心主广场。5.将公众参与的文化娱乐活动作统一考虑,体现动态的展示效果。6.入口大门与雕塑美术馆作重点设计,大门采用简练的手法,体现雕塑般的艺术氛围;美术馆沿地形错落布置,突出现代的弧形飘篷造型。

建筑面积:15000m²
Total Floor Area: 15000m²
设计时间:2000 年
Design Period: 2000

The World Sculpture Garden is one of the top three art gardens in china with a vast coverage area and scale of development. The theme of this project is to create a modern sculptural garden in natural setting theme.

Design considerations: 1. Planning focuses on a modern landscaping theme. 2. western-modern cultural features are imbibed into the design. 3. Emphasis on the importance of culture as well as the public participation to realize the design objective.

Design solutions: 1) on the topic of "friendship, peace and spring", varying linetypes and groupings are utilized to divide the space into different gardens with different themes. The result will emphasize the intention of spatial variation and vitality. 2) The contrast created by the large site and small sculptural elements provides the space with multiple layers and forms. 3) Landscape design gives primacy to vegetation when creating waterscape. 4) An intensive composition style is adopted to enhance the central plaza. 5) Recognizing the role of public participation as well as the activities of culture and entertainment to celebrate the area's dynamic atmosphere. 6) The main entrance leading to the Sculpture Gallery is treated in simple style to enhance artistic appreciation of the sculptures. The Art Gallery is arranged in a random order to emphasize the modern curved form of the sculptural canopy.

长春雕塑艺术馆
SCULPTURE ART GALLERY, CHANGCHUN

　　该项目位于长春市人民大街东侧长春市雕塑公园园区内，邻近主入口，位置优越，基地地势起伏，有6～7m的高差，景观资源丰富。

　　本项目以谦虚、雕塑感、象征性作为总体布局的原则，充分利用地形，形成人车分流的交通体系，车行道及货运在下层，人行则位于上层。

　　立面设计分为外、内二面，朝园外的西、南面以石材、木材形成壳状的整体雕塑感的外观，向园内的北、东二侧以玻璃、金属形成正常尺度通透的外观。馆外运用石材、木材、金属、玻璃的组合，都以浅色作为基调，强调简洁、精密。馆内则以白色为基调，将各种功能性装置尽量隐性布置，形成明快、整洁的背景效果。

建筑面积：12622m²
Total Floor Area：12622m²
设计时间：2001－2002年
Design Period：2001－2002
竣工时间：2004年
Completion Period：2004

The project is situated within Changchun Sculpture Park, near the main entrance. The project site has an excellent location with undulating contour of 6～7 m height level variation and an abundant landscape resource.

'The sculpture, modesty and symbolisms' are the general principles guiding overall arrangement. Optimum use of the landform is necessary to realize an efficient traffic system by setting the driveway and freight facilities on the lower floor while pedestrian traffic is managed on the upper level. The exterior facade employs a combination of stone, timber, metal and glass arranged according to the material colour tint to emphasize concision and precision. All non-aesthetic accouterments are disposed in a recessive way to reduce complexity and to mould the interior into a bright, orderly, and clean space. The south and west facades that face outside to the garden are treated with stone and timber while the north and east facades which face the interior are treated with glass and metal for an articulate appearance.

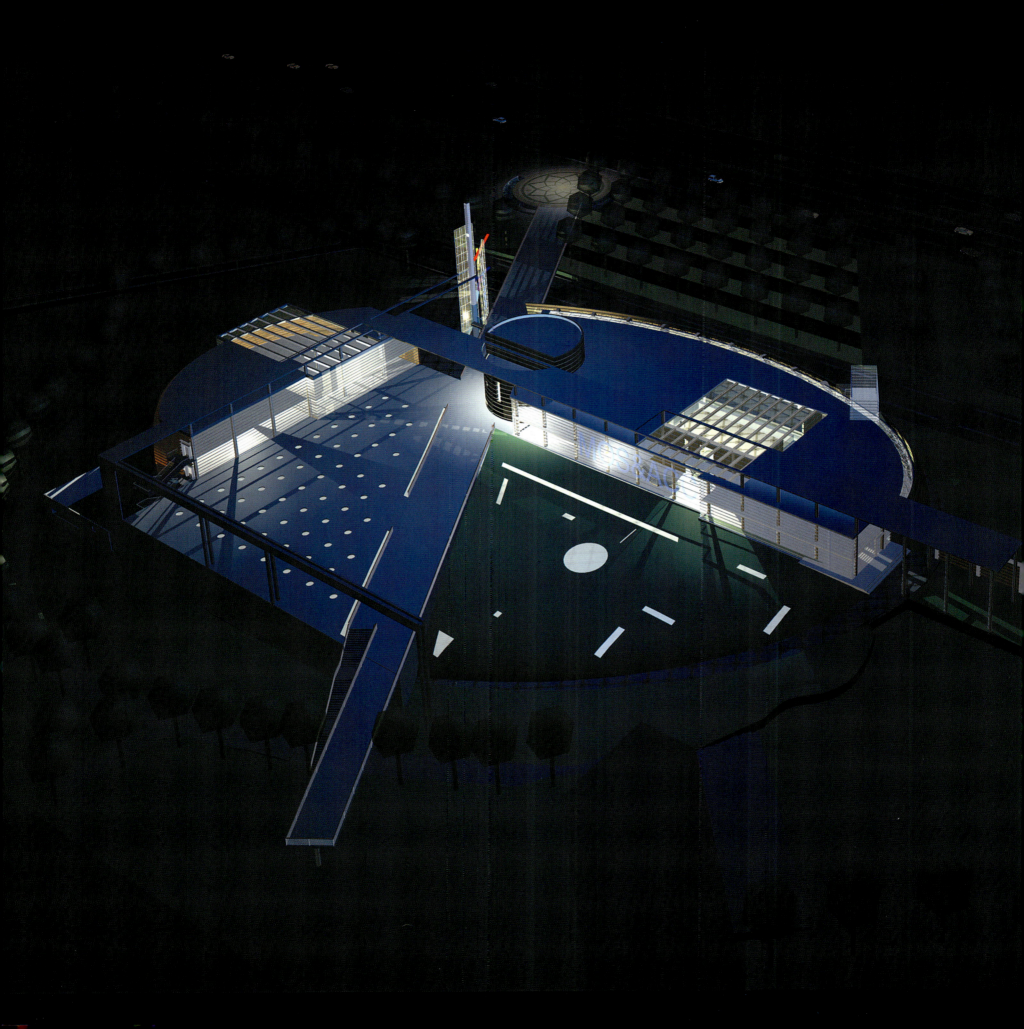

长春全民健身中心
SENIOR CITIZENS' FITNESS CENTER, CHANGCHUN

该项目位于长春市南岭体育中心北部,北临体北路,东面为繁华的亚泰大街,南面与五环体育馆相望。

本项目为一水平伸展的长条形建筑,建筑跨过区内道路而建,形成体育中心的北大门。在强调健身中心个性的同时,降低高度,减少对五环馆的遮挡,烘托五环馆、主体育场,以其作为整个体育中心北部的背景。通过曲、直体量的对比、直线与斜线的结合、自由的高低变化、形态丰富的外观造型以及活跃的内部空间,充分体现现代人对"运动"概念的理解。

入口广场设于北面,作为进入大楼及体育中心的缓冲空间,广场以硬质铺装为主,种植少量行道树及设主题雕塑。泳池东面、南面设庭院绿化,以草坪树木为景观特征。

建筑面积:15443m²
Total Floor Area:15443m²
设计时间:2002年
Design Period:2002

The project is located north of the Nanling fitness centre in Changchun. It sits adjacent to Tibei road in the north, Yatai street in east, and the olympic gymnasium in south.

The project is a strip construction constructed across the road, forming the north gate of the athletics centre. key points for consideration are: 1) creating an identity for the building project, 2) reducing the height and avoiding the obstruction, 3) deference to the Olympic gymnasium and the main gymnasium which are the main structures of the whole sport centre.

To accentuate full understanding to the modern concept of sporting through contrast between the curved and straight mass, the combination of straight lines and diagonal, height changes and other various treatments to create an active interior space.

The entrance square is located in the northern sector to function as a transition space leading to the main structure. Hard paving is adopted on the plaza ground surface, accessorized with some planting strips and sculptures. The swimming pool is defined by its surrounding courts and unique landscape.

深圳大学基础实验室工程
THE BASIC LABORATORY OF SHENZHEN UNIVERSITY

深圳大学基础实验室工程位于深大南校区,设计体现了大学是自我管理的学者社团。

通过对基地周围城市及校园环境的整体分析,实验楼在总体布局上试图最大限度地利用基地周边的优势并避免劣势的影响,沿北、西、南三向的周边布局模式延续了校园的景观轴线。围合的建筑群体体量朝东打开,将景观延伸进入建筑组团内部,不仅丰富了建筑组团内部的视觉空间,也使校园的景观主轴得到延展,从而更加完整。围合的院落向不利朝向封闭,向主要人流经过的景观朝向开敞。每幢建筑都因此获得最大的景观展开面,具有良好的日照及通风。

由于学生的人数多,流量大,人员密集,因此将学生实验室布置在建筑的下部一到六层。教师行政办公部分则布置在建筑的上部七到九层,利用垂直分区合理安排人员密度。通过在三层的位置布置架空层,将人活动的层面提高到了三层的位置上,相当于将一个高层化解为一个多层。

建筑面积:47000m²
Total Floor Area: 47000m²
设计时间:2006年
Design Period: 2006

The basic laboratory of shenzhen university is located in the new campus ,the design of it expressed our understanding of the University-the self-directed organization of the scholars.

The condition of the site defined the form of the building. The building unit the two parts of basic laboratory which have different fuctions into one complex .The complex lie around the north,west and south part of the site to form a big yard which extrude the landscape axis of the campus planning. The courtyard close to the unfavorable direction and open to the view direction which meet the main crowd get past. By doing this,each part of the building can achieve the largest open view as well as good sunlight and ventilation.

Due to the large number of the highly concentrated students, The building place the laboratory from the ground floor to the sixth floor and contains the majority of the office workstations on the upper three storeys. We arrange the people density by demarcating the vertical funtion region and at the same time we divide the high building into two lower ones with the placement of veranda floor.

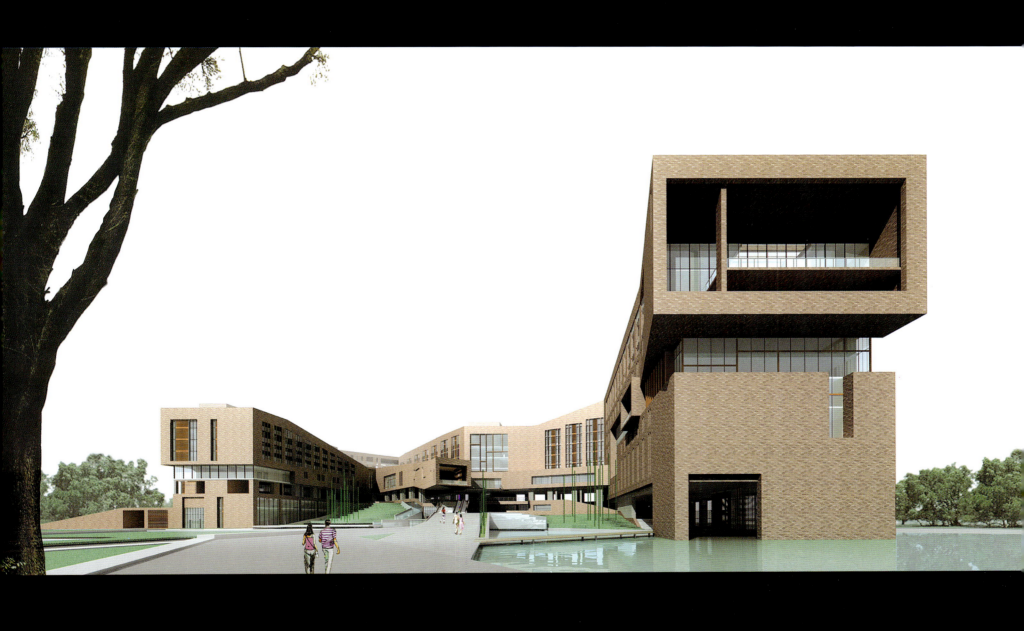

济南·综合广场
INTEGRATED PLAZA, JINAN

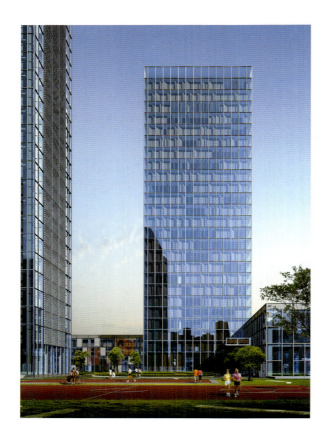

济南·综合广场项目坐落在济南名胜趵突泉以北，近临趵突泉公园、泉城广场，景观秀丽，视线良好。西侧有高架桥，噪声较大，北侧有学校，东侧有回民小学，基地南、北两条城市道路宽度较小，且属于旧区建筑。

方案传达一种力量，摆脱传统象征性的束缚，建立起济南新的形象和信心。从此，济南·综合广场不再简单是一组建筑，而是成为一个"地方"。整个概念是要构造一块地面，一个建设版图，一个空间发展城市，一个具有大型水平（中庭）的宛如高山般的地方。以供人们从这里看世界，并同世界交流。

建筑师按照南北与东西两个方向组织建筑。这样，就可以保证街区内部的道路能够齐整。另外，把建筑体量沿顺河东路后退，给城市提供一个慷慨大度的开放视角。所以，这个概念表现为一个直角形式的建筑，它面向东西方向敞开着，指向邻近的城市中心——趵突泉。每个入口都有独立的空间特性，简化了管理、简化了建筑组群交织的人流关系，增加了巨大的沿街面积。

每个建筑都通过一个特殊空间被连接到其他建筑，该空间叫做"水平中庭"。

这个交界面是在多个层次上的。这些中庭都是透明的、开敞的。自然光可以透过建筑，并打破直角立面的特性，这些光线为建筑带来了一个真正的自由。同时，在夜里，这些水平中庭都活跃着光照，让城市中的人们可以清楚地看到它。事实上，它们在建筑的内部构成了新的"城市空间"，方向指向城市的主体。正是通过这种方式，人的尺度得以在这组宏伟的组群中调和、识别。

建筑面积：143250m²
Total Floor Area：143250m²
设计时间：2006年
Design Period：2006

The Yinzuo Jindu project is located north of Baotu Spring, which is one of the famous sights of Jinan city for its beautiful landscaping and natural scenery.

The concept aims to reduce symbolic restrictions of tradition to create a new image and attitude for Tsian. Therefore a 'space' is proposed, instead of a single building. It is assumed as a community, a developing urban space, a large-scale open area to examine and communicate with the universe.

An orthogonal arrangement is applied to the building groups to ensure a clear and orderly network system. In addition, the whole building group is setback so that a munificent and open viewing angle can be availed of. The isolated space of each entrance, with its own distinctive look, simplifies the management system and pedestrian flow within the complex while enlarging the street area. Every building is linked to another through a transparent and open atrium to enable better communications between the enclosed spaces. Furthermore, these atriums also break down the old conventions of an orthogonal elevation as it conveys more daylight into the buildings, for energy conservation. At night, the atriums create illumination for the urban space to reveal identification and harmony of the human scale with this grand complex.

西安高新区 CBD 商务公寓楼
HI-TECH CBD COMMERCIAL APARTMENT, XI'AN

项目位于西安高新区二次创业规划区，属于未来西安高新区的核心区域，占地面积29229m²，拟建一集商务办公、居住、SOHO、商业、休闲、娱乐、餐饮、运动等多种功能于一身的现代城市综合体，以期实现最佳的用地潜力，并为此区域规划的开拓奠定良好的基础。

设计将项目理解为高科技新区一场典型的"拓荒运动"，针对新区用地充足、人气弱、资源少、地势平坦、空旷开阔、可塑性强的特点，给出了大尺度、综合性高、配套丰富、辐射力强、符合总体规划思路的、引导性较强的优化设计，功能上以解决问题的实效性为基准点，外观上强调尺度的协调与对比以提升对区域的空间影响力。此外还注重对经济耐久性、节能环保、功能可变性等的考虑。设计充分利用了基地内的通道条件和西面公交广场的预期作用，对新的城市形态和大众行为模式进行了"预设"，营造出一幕可以化为现实的新荒中的"海市蜃楼"。

建筑面积：93248m²
Total Floor Area：93248m²
设计时间：2006 年
Design Period：2006

The structure which covers an area of 29229m², will be the core of the Xi'an Hi-Tech district in the future. It serves as a modern urban integration combining offices, residences, business, leisure, entertainment, dining, and sports. The aim is to liberate the potentials of the site and to carve out a foundation for future urban planning in this district.

Building on site features of abundant fields, sparse population, lack of resources, flat topography, open and flexible spaces, the design is comprehensive as it is massive. It is large-scale, highly integrated, flexible, with comprehensive support facilities, intended to turn idle land into a delightful space with value, sustainability, economy and flexibility. Through full utilization of existing roadways and a conventional public bus square, the design realizes the vision of a renewed urban configuration and an improved lifestyle.

三亚中油大酒店
SANYA ZHONGYOU GRAND HOTEL

三亚中油大酒店位于海南省三亚市三亚湾。基地北面群山环抱,南面与海相邻。视野开阔,景致优美,地理位置得天独厚。

该酒店为国际标准四星级酒店。酒店首层及二层为公共设施,包括各种餐饮、休闲空间以及服务用房;地下层为设备用房及内部职工用房;三层至八层为客房层;九层为行政客房层,客房层可通过南侧园林观赏到海景。

作为酒店主体的客房楼设置于用地中部偏北的位置,南侧面向大海,客房楼折线形的设计,能争取更多的面海房间。客房大楼将用地划分为南北两个园林,北侧园林营造入口的迎宾气氛,并使建筑远离城市道路;南侧园林为客房及公用设施提供大面积的自然景观。为充分获得自然景观,公共设施部分向南延伸至南侧园林之中,客人使用公共设施的同时,将享受到室外自然景观,良好的采光通风,便于营造各具特色的空间环境。

建筑面积:25000m²
Total Floor Area: 25000m²
设计时间:2006 年
Design Period: 2006

The construction sits in the Bay of Sanya, Sanya city, Hainan province. It is surrounded by mountains in the north, and neighboured on ocean in the south, with a wide vision, beautiful sight, and splendid position.

It is a four-star hotel. The first two floors are public facilities including various rooms for dinning, leisure and service; the underground is the room for equipment and employee; guest rooms are distributed on the floors of 3-8; the ninth is the administrative rooms through which the sight of the ocean can be seen.

The main building for guest rooms adopted folded style to get more rooms which face the ocean. The guest-room building divided the plot into two gardens, the north one create the atmosphere of welcoming the guests and free from the urban road; the south one offered the natural sights for the guest rooms and public facility. The public facilities extended south to south garden to bring more natural sight and good lighting and aeration to create the special space environment.

绵阳安县罗浮山温泉山庄
LUOFU MOUNTAIN HOT SPRING VILLA, MIANYANG AN COUNTY

该项目位于四川安县桑枣镇西南之罗浮山温泉度假区内，是一个为四川绵阳地区举行会议、休闲、度假的多功能旅游设施。

温泉山庄基地位于温泉度假区范围之最高处，基地为多边形平面，四面环山，坐北向南。该设计总体布局的主体思想是：依据地形、地貌，使体形大小各异的建筑与之有机结合，形成一组依山就势、高低错落的建筑群体；尽最大可能保护基地原有4片较大林木；充分利用原有山间水塘，并进一步适当拓展，在基地主要区域形成较大水面，将建筑群体融于优美的自然环境之中。

整个建筑群均采用中国传统民居的坡屋顶造型，并以此表达山地建筑的特色。由于大面积采用了天然建筑材料，如毛石墙、涂料等，使建筑形象纯朴、大方、色调和谐，并具备丰富的地方特色和文化内涵，达到了一种诗意的境界。

建筑面积：32065m²
Total Floor Area: 32065m²
设计时间：2000年
Design Period: 2000
竣工时间：2002年
Completion Period: 2002

The project is located in the hot spring holiday district southeast of Sangzao town, Mianyang An city, in Sichuan. It is a multi-functional resort combining business, leisure and accommodation.

The site of the hot spring villa is situated on the top of the holiday district, in a polygon shaped lot that is surrounded by a mountain range. The main principle of the overall layout is to create a building complex sited along the mountain side and set with varying heights that matches the landform and physiognomy while protecting its pristine setting. To fully utilize existing waterscape, a large water surface is formed to expand space and develop harmony of architecture and nature.

Traditional sloped roofing is used for the whole structure to further express distinct qualities of mountain architecture. Treatment of raw construction materials such as rough stonewalling and coating in large areas make the construction appear rustic, grand, and harmonious, while it showcases local specialities and cultures.

南京审计学院
AUDITING ACADEMY, NANJING

该学院由莫愁、浦口二校区组成，新校区毗邻浦口老校区而建，占地约113.32万㎡。

本项目将山脊作为公园的绿带与人工湖组成的蓝带并置，形成以中心大公园为核心的生态型校园布局，并以此作为大公园系统的核心与起点，将各用地以公园"链"串联而又间隔开来，形成独具特色的校区公共景观体系。山、水作为主题的大公园为整个校区注入了强大的活力并成为活动的平台，为严肃的校园生活带来必不可少的谐和。

新校区的环境自然生动，起伏有序，适应审计学院的特点，以清晰的几何秩序来统率整个校区的场地结构，礼仪、教育、文体三条主轴交会于主广场，学校五大分区有序地组织在一起，形成动中有静、静中有动的格局，并将图书馆作为主地标以控制全局。核心、教学、文体、学生居住、教工居住五大分区明确清晰，既方便联系，又互不干扰，提高了校园的管理和维护的效率。

建筑面积：266900㎡
Total Floor Area: 266900㎡
设计时间：2003年
Design Period: 2003

Consisting of Mochou and Pokou campuses, the proposed campus covers an area of 1133200m² and sits adjacent to the old campus.

The juxtaposition of a green belt- natural environment and a blue belt - a man-made lake, forms the ecological campus. It is then taken as the core area and central node connecting each of the functional spaces to create an ecological campus.

The core, which is based on a water and mountain theme, injects vitality and freshness into the project to break through the regimented atmosphere of conventional campus life. The new campus environment is natural, lively, fluctuant, and highly appropriate to its academic purpose. The campus structure is restrained by a clear geometric order with three principal axes of amenity representing education, culture, and physical well being converging at the main square. The campus is divided into five districts organized together to form a pattern of activity and calm. A clear boundary between each district offers a convenient connection, effective segregation, and a system for efficient campus management.

深圳国家工商行政管理总局学院
STATE INDUSTRIAL & COMMERCIAL ADMINISTRATIVE BUREAU INSTITUTE, SHENZHEN

该项目位于深圳南山区，龙珠大道旁，与桃源村隔路相望。基地内地形高差变化复杂。

本项目由行政教学楼、图书馆、文体中心、公寓楼、食堂、后勤等几部分组成。结合地形高差变化，设计将基地分为三个台地，将建筑按不同的功能进行整合。第一台地为教学区，第二台地为学员与教职工的生活及运动区，第三台地为专家居住区。建筑布局与环境形态相结合，围合中心主楼，构成"双龙戏珠"的形态，并顺应山势形成中心跌落大花园。

建筑造型突出了不同功能建筑的特点，同时又达到了和谐。行政楼形态意象取自中国古代建筑"明堂避雍"的形制，造型有力而大方；图书馆采用与山地结合的生态造型；文体中心充满动感；公寓楼则盘山而建，形成错落有致的山地建筑形象。

建筑面积：32000m²
Total Floor Area: 32000m²

设计时间：2004 年
Design Period: 2004

The project site is situated in Nanshan district, Shenzhen. It has a complex landform where it lies along Longzhu Avenue, adjacent to the Taoyuan Village.

This project is a multi-function complex consisting of an administrative teaching building, a library, a culture and fitness centre, a large dining-hall, a logistics centre and an apartment block. In response to its fluctuant landform, the site is divided in three mesas arranged logically according to the nature of its different functions. The first mesa is the teaching area. The second mesa is residential with a fitness area for students and staff and the third mesa is a residential area for specialists. The layout of this building complex is corresponds with its natural setting forming an enclosed central space where the main building is located.

The unique structural appearance enhances the functions of different buildings as well as it harmonizes the whole building complex. The form of the administrative building denotes ancient Chinese construction with a powerful image and an appealing shape. The ecological theme of the library, the culture and fitness centre, and the apartment comes from the mountain, representing a remarkable image of mountain architecture with a random but orderly style.

南京中国药科大学——江宁校区
SCHOOL OF TRADITIONAL CHINESE MEDICINE, JIANGNING BRANCH, NANJING

该项目位于江宁大学城东南部，东北毗邻宁杭高速公路，西靠前进河，北依104国道和南京二环路等主干道，西侧遥望方山美景，校园南侧和西侧为60m宽景观大道。

整个校区从东至西竖向分为科研区、教学区与行政区、生活运动区和家属区四大功能区域，中间间隔着三条生态系统。各功能组件形成特定的建筑单元，通过各种灰空间，形成一个链状序列。

行政区与校前区是学校对外的形象，由一幢行政主楼与礼堂会议中心、接待中心共同形成了对校前广场的半围合，建筑风格庄重、大方、现代。

在生态地毯的概念之下，在与药用植物园连接的同时，联系南北生态绿地形成一个生态整体。以自然山水绿化为主，人工小品为辅营造生态自然的环境，折线形的建筑形态有利于对环境的充分利用，与自然有机结合。

建筑面积：675480m²
Total Floor Area：675480m²
设计时间：2003年
Design Period：2003

The project is situated in the south-eastern part of Jiangning University city. To its northeast is Ninghang highway, west are the Qianjin River and the Fang Mountain, north is a national highway and the Nanjing second ring road, and to its south is a 60m wide sightseeing strip.

The campus is divided into four functional districts of scientific research, teaching, administration, and living and sport linked by three ecosystems. Each functional component forms a special construction unit and becomes a link in a chain sequence. The administration district and the front of the campus make up the image of the university, consisting of the main administration building, auditorium, meeting centre, and reception centre, so this block is designed in a grand, decent and modern style.

On the issue of ecology, a landscape system is shaped to integrate with the medicine arboretum and to connect with greenbelts in the north and south sides. Emphasizing primacy on natural waterscape and landscape, the design adopts the folded architectural form to enhance an intimate relation with the environment and its organic cooperation with nature.

居住建筑
Residence Building

深圳星河国际
XINGHE INTERNATIONAL GARDEN, SHENZHEN

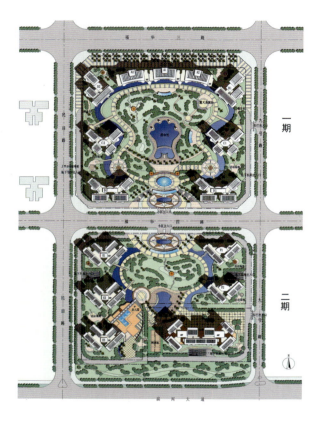

该项目位于深圳市中心区南部,北面为"购物公园",东侧为"深圳市会展中心"。

整个项目分两期开发,功能为大型商业和高档住宅。
整体规划:小区的环境空间作为中心区生态空间的子系统。
环境设计:住宅底部架空,使中心花园与架空层绿化相渗透,形成开放的小区内部环境,突出"生态大花园"的主题。
住宅设计:户户均有良好的朝向与景观,每户均保证自然穿堂风。
造型设计:突出中心区城市住宅的品质。

建筑面积:265184m²
Total Floor Area: 265184m²
设计时间:1999－2002年
Design Period: 1999－2002
竣工时间:2003年
Completion Period: 2003

获奖情况 Awards:
2004年深圳市第十一届优秀工程设计一等奖
first prize,"11th Excellent Building Design Award" for shenzhen, 2004
2005年广东省第十二次优秀工程设计一等奖
first prize,"12th Excellent Building Design Award" for Guangdong, 2005

The project is situated in the south of central Shenzhen, to its north is a shopping district and to its east is the Shenzhen Exhibition Centre.

The whole project contains two phases, providing functions for a large-sized commercial quarter and a high-class residential quarter. The environment of the residential quarter is created as a subsystem of the central ecological space. The theme of 'a huge ecological garden' is implied by inserting overhead spaces on the ground floor which will enhance infiltration between the central garden and the overhead spaces. The elevation design strives to reveal a central urban housing estate with significantly improved qualities. Its rational layout guarantees excellent orientation, scenery and natural ventilation for each resident.

南京天泓山庄
TIANHONG VILLA, NANJING

该项目位于南京东郊，紧邻紫金山风景区。规划用地面积205500m²。由16栋高层、中高层，37栋多层住宅以及会所、幼儿园组成。

立面划分采用三段式的经典划分方式，低层部分采用粗犷的毛石材料，以强调建筑的基座感以及和环境山地的融合。主体的中部采用现代主义手法，简洁明快，屋顶构架结合屋顶退台及屋顶花园做建筑化的处理，是塔楼部分设计语言的理性的延伸及收顶，极大地丰富了塔楼顶部的空间层次。

塔楼部分用凸窗、转角窗、落地窗的设计手法，运用竖向和横向板带线条将阳台凸窗连成一个整体。主墙面为白色调，错落和凹凸的墙面为浅黄色，加上基座的石材贴面，对比中保持协调，有层次感而不失整体，创造了平和亲近的立面风格。

会所采用虚实对比的手法，透明生动的玻璃幕墙与金属板挑檐相互对应，展现出建筑轻盈通透的特征。

建筑面积：2435557m²
Total Floor Area：2435557m²
设计时间：2003 — 2004 年
Design Period：2003 — 2004

The site is situated east of Nanjing, abutting the Zijin Mountian landscape. It covers an area of 205500m², consisting of multi- level residential structures, a club, and a kindergarten.

A Classical three-section elevation design is used as the general treatment for residential buildings. Raw stone is used to stress the base and to facilitate connection with the environment. The middle part is treated plain, bright, and simple in accordance to its modernist theme. Structural termination at the top is an extension of the design process, thereby enriching the spaces on top. The design of different windows connects the balcony and the bay windows as a whole. White main wall surfaces, light yellow walls, and stone surfaces creates synthesis for the varying wall treatments and thus, generating an image of a solid and secure, yet cozy and gentle elevation.

A Combination of voids and solids and of glass and metal at the club structure creates a surprisingly rich appearance for a light structure.

深圳香域中央
XIANGYU CENTER GARDEN, SHENZHEN

该项目位于农科开发区北面——农园路和莲花西路交界处，地块东、东南与深圳市大型植物公园接壤。地理位置极其优越，是高尚、典雅的理想居住社区。

设计注重了以下要点：

1. 充分利用地下、地上、建筑屋面、退台等空间，通过各种手段为居民提供休闲、娱乐、停车等各种类型的户内外生活环境。

2. 采用"三维立体"绿化手法，让绿地、庭园有高低起伏的变化，同时将绿地引入建筑，实现小区全方位的绿化，特别是将小区东侧植物公园绿化引入，使本社区自然成为植物公园的一部分。

3. 强调了住户的均好性原则，使每户都拥有良好的朝向，优美的景观。

4. 曲线与直线相配合，点式与板式住宅相映衬，体形高低错落，外部空间更趋丰富。

建筑面积：185331m²
Total Floor Area：185331m²
设计时间：2003 — 2004 年
Design Period：2003 — 2004
竣工时间：2005 年
Completion Period：2005

The project is situated north of the Agricultural Science Development Zone, at the crossing of Nongyuan and Lianhua roads. The site is bounded on its south and east by a public park. Its excellent location dictates its position as an upscale residential district.

The following key points are major design considerations:

1. Optimal use of surface ground, basements, rear yards and a careful layout offers ideal living amenities both indoors and outdoors.
2. Three dimensional landscaping created by green areas, yards, and surface variations conveys green areas into built structures.
3. Upscale residential design, proper orientation, and excellent views are guaranteed for each residential space.
4. Surface treatments of curves and lines, textures, and the interaction of varying shapes and heights to create a rich elevation design.

澳门环宇天下
HEISHAHUAN MIDDLE STREET BLOCK R+R1, MACAU

该项目总用地面积为13917m²。其北连珠海关闸；南接澳门国际机场及澳港码头，对外交通非常便利。

本项目以大海为背景，以东南亚风情园林为元素，演绎出澳门独一无二的主题式园林及会所休闲区。并利用空间创造出特有的空中花园，为住户提供更多样的活动空间。

商业裙楼注重整体性及灵活性，立面设计上突出了标志性及统一性，在人流密集的南端设计了一标志性圆形商业空间，内有电梯厅直达花园平台层。商铺沿街立面统一考虑设置遮阳装置及空调机位和招牌位，商铺内预留高窗，方便日后改建成阁楼供办公或货仓使用。

户型设计追求室内外空间的和谐，充分满足住户生活的需求，户型间均不产生对视，户户均有良好的采光通风。

建筑面积：160000m²
Total Floor Area: 160000m²
设计时间：2004年
Design Period: 2004

The project site covers an area of 13917m². To its north is the city of Zhuhai and to its south is the Macao International Airport and the Macao port.

Using the ocean as background theme, and the gardens of Southeast Asia for inspiration, the project achieves a unique themed garden and leisure district. Special, elevated gardens from space offer an alternative space for multiple activities.

The commercial area exhibits unity and flexibility. Signages are extruded on a vertical plane, and a commercial space with lift going to the elevated garden level is located in the south.

The design of the exterior pursues harmony between spaces, and satisfies the requirements of the residents. Each residence has the good, unobstructed orientation and ventilation.

长春中海水岸春城——莱茵东郡
CHINA OVERSEAS RHINE GARDEN, CHANGCHUN

用地位于吉林省长春市静月开发区卫星路南商住区,总建筑面积32.9m²。

规划设计采用内环模式,带状布局;水景自然区形成6个"别墅半岛",浑然天成。半岛别墅群镶嵌在以森水景观带为主体的开放空间中,营造出"林语水岸别墅公园"。完善的步行系统提升了公共景观利用率。

景观设计的核心是1.2万m²的森水景观带,传承自东面一期歌雅园的水景轴线,往南延续到南地块的开发,完成三地块"大景观—大水系"的整合设计。设计尊重长春气候特点,注意绿化和水景结合,控制水景面积,节约成本。

住宅产品有三大系列。立面设计力求多元化的同时注重统一,对空间节点处的住宅进行特别处理,增加尖顶、高塔、退台等元素,增强识别性;外立面构件以精细划分和小尺度构件达到亲切近人的效果。秉承了一脉相承的欧式经典大宅风格,大量采用简化欧式符号作为细部,应用现代建筑材料,增加时代气息,展现原汁原味的地道欧式小镇风情。

整个设计形成"一个别墅公园,五个特色组团",依托临河资源,强化水景优势,设计德式经典立面,打造莱茵临水小镇风情。达到了规划、建筑与景观浑然一体,和谐天成的效果。

总建筑面积:329000m²
Total Floor Area: 329000m²
设计时间:2005年
Design Period: 2005

The project which has a 268445m² area is situated in a commercial district on Weixing road, Jingyue development district, Changchun city, province of Jilin.

Structures are sited within the site's inner ring and are laid out strip-like. Six villa peninsulas of varying sizes are created by shaping the landscape into cove-like spaces along its fringes to create focus on the waterscape. A landscaped sightseeing strip, 1200m² traverses the district to optimize views

Coordination and unification are planning priorities while all other discrete elements are considered. Rigorous detailing is employed to particular nodes (such as residences along the streets) to heighten particular elements (e.g. spires, towers, steps, etc.) resulting in enhanced massing and focal points. An adaptation of grand European houses and usage of modern construction materials and techniques accentuates the charm of an ideal European town.

Finally, the project forms a park villa comprised of 5 distinct communities, each with its own special attributes, creating an effect of complimentary individuality that gives the area a collective atmosphere of freshness and dynamism.

苏州中海·半岛华府
CHINA OVERSEAS ROYAL PENINSULA, SUZHOU

该项目隶属于苏州工业园区，位于园区金鸡湖的金姬墩住宅区东侧，用地北侧为金鸡湖，南侧为规划路，西侧为规划河道及绿化，东侧星湖街与本地块之间规划为15m宽城市绿化带。

项目定位为苏州高档精品人文国际居住社区，通过适度创新领先市场，通过欧式风格的建筑与园林景观的塑造，充分利用地块周边优美的自然环境，塑造低密度联排别墅（TOWNHOUSE）社区。

园林设计风格与小区建筑风格相协调。总体考虑欧式特点、意大利风情。A地块由主入口进入，在退后城市绿化带位置设置铺砌广场，设计小区主题景观。

B地块结合小区主路两侧景观，进行局部节点景观设计，不但改善了两侧联排别墅的景观条件，亦使车行环境改善，通过中线景观轴的设置，将中部户型的价值提升，并形成相应的步行区域。

户型设计充分利用景观和朝向资源，整个小区所有户型均为南北通透。联排别墅端头户型保证三面采光和景观，中间户型通过中空庭院和前庭后院，使通风采光得以优化。二期别墅还按每栋合院设计，停车分别南北进入，使两栋间围合共享生活空间，使前庭、中庭、后院、公共庭院空间串联，增加户型的景观和人文附加价值。A区两栋多层住宅为创新多层户型，于北侧设置空中大露台，增加住户室外景观空间，尽享金鸡湖美好景观，户型方正实用、南北通透。

建筑面积：69487.8m²
Total Floor Area: 69487.8m²
一期建筑面积：37328.50m²
Covered Area in I: 37328.50m²
二期建筑面积：32159.30m²
Covered Area in II: 32159.30m²

The project is proposed as an international residential community. Design innovations lush vegetation gives character to this low- density townhouse community.

The design theme adopts European features with Italian influence.

Sight nodes on both sides of the main road in block B improves, not only the views towards the townhouses, but of the environment as well. A sight axis built on the middle of the site adds value to the residences in the centre and provides an ideal area for walking.

Site layout offers each residence with optimal orientation. The design not only ensures the views of each residence, but it also optimizes ventilation and lighting. For residences in district A, large terraces were provided to enlarge their viewing spaces

The design of phase II forms a continuous space of open spaces, yards, etc., thereby increasing the area's visual treats and enhancing its general atmosphere.

江门中天国际
ZHONGTIAN INTERNATIONAL GARDEN, JIANGMEN

该项目位于江门市新区中轴线上,白石大道北侧。地块呈凹形,被道路从中间分隔成两块。由于项目所处位置的重要性,本小区的定位是CBD内具有国际品位、现代风格并与地位特点相适应的后小康智能化生态型住宅,其商业是集购物、餐饮、休闲娱乐于一体的商业群。项目结合南方特点,充分利用自然条件,营造富有特色的水系空间,创造"以人为本"的人居环境。

在规划布局上考虑到江门具体的气候特点,建筑以南北布置为主,点板结合的布局有利于夏季通风及日照要求。

在绿化系统的设计上,以一贯穿南北的线形绿化空间,大手笔地将东西两块基地有机地联系起来,形成了"一条轴线、二个层面、三种空间"的多层次绿化空间。以水系作为组织绿化的主要元素,营造了大量的亲水空间,有利于区内小环境的调节,较高的绿化率亦进一步体现了人与自然共生的理念。

建筑面积:394300m²
Total Floor Area: 394300m²
设计时间:2003年
Design Period: 2003
竣工时间:2005年
Completion Period: 2005

The garden lies on the middle axis of a new district in Jiangmen, in the northern part of Baishi road. The concave shaped site is divided into two plots by a road. Due to its important location, the project is conceptualized as a well-planned ecological district of international standards with a modern theme. Its commercial sector is an aggregation of shops, restaurants, services, and general leisure and entertainment facilities. In consideration of the specialities of the south, and to optimize use of inherent natural resources, a special waterscape is created to amplify interaction between man and his environment.

In consideration of the special climate of Jiangmen, structures are oriented along the north-south bearing to satisfy ventilation and sunlight requirements especially during summer season. A multilayer green space created from "one axis, two layouts and three spaces" adds a special feature to its general landscaping. Using water feature as the main element for landscaping to create the abundant near-water spaces is an efficient way of improving the environment for the district. The project's high green-area ratio affirms the concept's concomitance between man and nature.

深圳水榭花都三期
SHUIXIEHUADU PHASE III, SHENZHEN

该项目位于深圳市香蜜湖片区,香梅路与红荔西路交会处。由六栋高层(28~29层)住宅,和一栋二层商业市场组成。建筑沿用地周边环抱布置,在三期与一、二期之间形成大片的园林绿化空间。

本项目延续了一、二期简洁明快的建筑风格,同时又力求有特点,在统一中寻求变化。采用形体对比,竖向线条与横向阳台飘板相结合,突出挺拔俊逸的外形。墙身结合凸窗与百叶等金属构件组合,使层次更加丰富。所有管线及空调机位置在立面上均设计了隐蔽措施。空透的大阳台,大面的玻璃窗与浅色的框架和墙面形成虚实的对比,突出了南方热带建筑自由、灵活、通透的特征。

建筑面积:127807.53m²
Total Floor Area: 127807.53m²
设计时间:2003 — 2004 年
Design Period: 2003 — 2004

It is situated on the crossing of Xiangmei road and Hongli west road in Xiangmihu, Shenzhen. The project consists of by six high-rise residences (28~29 floors) and two floors of commercial space. Building structures are dispersed along lotlines to form a large central garden together with phases I and II.

The design theme of simplicity and brightness employed in phases I and II is sustained to achieve unification. Its main shape is enhanced by the coordination of subordinate of shapes and the combination of vertical lines and a transverse board. Balconies, large glass windows, and light framing enhance the specialities of a vibrant, Southern lifestyle.

深圳华侨城锦绣花园三期
OVERSEAS CHINESE TOWN JINGXIU GARDEN PHASE III, SHENZHEN

该项目位于深圳市华侨城东部，深南大道以北，华侨城东路以西。

项目邻近深圳湾、锦绣中华、红树林鸟类自然保护区、民俗文化村等著名旅游胜地。地理位置优越，交通极为便利，功能为高尚住宅。本设计与一、二期良好融合形成完整的总体环境，以弹性的围合形成尺度较大的花园。单体设计中采用开放的公共交通空间设计模式，所有楼梯间均直接对外开窗，公共走道、电梯厅为开敞式设计。每三层一处设有近百平方米的空中花园，使得电梯厅成为邻里交往的半私密场所。住宅阳台外侧设置有活动百叶，起到遮阳及遮蔽不良视线的效果。阳台百叶每家开启不同，阳台每几层规律变化，形成了丰富、整体而又统一的建筑立面肌理。

建筑面积：114733m²
Total Floor Area：114733m²
设计时间：2000 — 2001 年
Design Period：2000 — 2001
竣工时间：2003 年
Completion Period：2003

The project is proposed as an upscale residential project. It lies in the heart of Huaquiang district, adjacent to famous tourist spots, with an excellent geographical location and convenient traffic access.

The combination of phases I and II encloses a complete, self-contained community where the enclosure also shapes the central garden. An open public space is used in each residential structure where a hanging garden is created on the third level to create space for an atrium above the lift hall area. Louvers outside the balcony functions as additional shelter and changes made on intermittent floor levels sculpts the building shape into a richer elevation design.

利群连云港住宅小区
LIQUN LIANYUNGANG RESIDENTIAL DISTRICT

本项目位于连云港市行政中心区西侧，总建筑面积约55万m^2，容积率2.20。用地周边景观资源优越，东侧是280m宽城市绿带，南侧为拥有宽大湖面的城市公园。

通过分析周边现有资源，规划一条"森水景观带"，把东侧绿化带和南侧公园有机整合；四级规划结构凸显"泛公园"规划新理念，人性化院落与景观带交融共生；建筑群体空间呈现"北高南低"的关系，较自然地完成了城市向公园的空间过渡；三级景观结构开合有致，建筑群体空间婉转流动，营造动感、活力之城；住宅产品以高层、中高层、多层和4层情景洋房为主。

整个楼盘从规划、景观、空间、单体等多方面精心设计，使其成为连云港的一个顶级标志性超大型社区。

建筑面积：550000m^2
Total Floor Area: 550000m^2
设计时间：2005年
Design Period: 2005

The project which has 550000m^2 covered area and a 2.20 plot ratio is located west of the Lianyungang administrative centre. Natural landscape within the area is excellent, with a 280m urban green belt strip to the east and a city park with a lake to the south.

A forest-lake view corridor is planned to conform with the green belt and park. The site plan proposes the concept of an extensive park to create harmonious interconnection between humanity and his environment. A "low in south, high in north" building configuration realizes the transition of a built city to a natural park. Interacting with the existing landscape, the sight creates a active and energetic city and the residences are distributed to well-oriented, multi level housing units.

Elaborate elevation design with strong emphasis on orientation, sightlines, and space planning sets a new standard for ideal city living.

成都云岭高尔夫别墅
YUNLING GOLF VILLA, CHENGDU

本项目位于成都双流国际机场附近，西邻川西平原，南邻成都国际高尔夫球场，区位条件优越，视野开阔，为纯别墅高档社区。本项目一期建筑面积近3万m²，包括独栋别墅、高档度假公寓、小区会所。建筑结合周边环境和基地现状，使建筑融入周边环境，突显建筑的豪华、大气、浑然天成。

本项目建筑风格采取美式别墅风格，突出平面功能的合理，与立面细部相结合，创造出一种具有休闲、家居、豪华的居住氛围。景观设计采用高尔夫主题，在景区内设有推杆练习场，同时由于小区紧邻成都高尔夫球场球道，与高尔夫实现零距离。

建筑面积：32000m²
Total Floor Area：32000m²
设计时间：2005 年
Design Period：2005

The project is adjacent to Chengdu International Airport, sharing boundaries with Chuangxi Plain to the west, and Chengdu International Golf Club to the south. Its excellent location provides distinctive advantages for a luxury villa district. The first phrase of the project occupies an area of 30,000 m² containing detached houses, luxury apartments and a club house.

The american villa theme is adopted, integrating efficiently with its natural environment and existing conditions. Rational organization and elevation details present a living atmosphere of relaxation, comfort and luxury. Conceptualised along a golfing theme, a series of putting greens has been provided to accentuate the feeling of life within the fairways.

深圳中航格澜阳光花园
ZHONGHANGGELAN SUNSHINE GARDEN, SHENZHEN

该项目位于深圳市宝安区观澜镇新中心区，东临大和路，西临沿河路，北为工业大道和观澜世纪广场。

本项目通过一个社区小公园将商业和住宅两大部分分隔开来。同时，以世纪广场及文化艺术中心为起点，形成一道贯穿商业及居住两部分的主轴，将二者又联为一体，并使该主轴可以成为将来南边地块的延伸主轴。商业区在用地北部沿大和路及工业大道设置，采用骑楼及平台等方式，将二层商业空间也联成一体，形成丰富的多层次商业环境。休闲中心设于近住宅区部分，使之同时能更好地服务于住宅区的居民。幼儿园顺沿河路独立设置，可单独对外开放。

立面通过对屋顶、廊架、阳台、窗套等元素的打造，引入丰富多彩的异域风情．结合色彩、材质的变化，营造一个富有情趣的商业空间和浪漫温馨的居住环境。

建筑面积：153081.59m²
Total Floor Area: 153081.59m²
设计时间：2004－2005 年
Design Period: 2004－2005

The project is situated in the centre of Guanlan Town, Baoan district, in Shenzhen. It is bounded on its east by Dahe road, on its west by Yanhe road, and on its north by Industry road and Guanlan Century Square.

A community park separates the commercial area from the residences. An axis through the two nodes formed by Century Square and the Culture Centre binds these open spaces into one integrated whole yet expandable axis in anticipation of future expansion. The commercial area is set in the northern side along Dahe and Industry roads. Its nested flat roofing unites two levels of commercial space into one and projects a multi layered, dynamic commercial environment. The leisure centre is located near the residential area to serve its residents better. The kindergarten, which is set along Yanhe road, is autonomous so it can be opened to the outside and function independently.

The combination of foreign influence and elegant treatment of pertinent structural aspects such as the roof, hallways, terraces and windows and the colour and material variations collectively create a residential atmosphere that is lively and upscale, as it is romantic and elegant.

合肥安高城市天地
ANGAO CITY GARDEN, HEFEI

该项目位于合肥市合作化路及望江西路交会处,毗邻中国科技大学。方案以景观先行为原则,400m长绿化水景步道连通南北两个景观节点,形成"一轴二心"模式,沿景观步道设大型中心绿化,产生震撼性空间效果。住宅组团围绕景观布置,西侧为由多层及中高层组成的绿荫水庭,庭院与水景形成互相交融的空间效果,最大限度地运用了水景。立面风格追求自然 摒弃多余的装饰,回归建筑的本质,体现居住与自然良好的关系,阳台在立面上成为构图元素,强调节奏感。住宅轮廓富于变化,中高层及多层采用退台形式,强调建筑体块间相互咬合穿插,变化丰富,体现诗意的设计内涵。

建筑面积:290000m²
Total Floor Area 290000m²
设计时间:2004年
Design Period:2004

The site is situated at the junction of Hezuohua road and Wangjiang west road in Hefei, adjacent to the China Science and Technology University.

Key point for this project is its scenery — a 400m² landscaped area and waterscape. Footpaths connect the sight nodes to one another to form "an axis with two ends". The large central landscaping bound by the footpath generates strong spatial effect of openness and residences are set along this path to generate access to the scenery. To the west is the where multi-level residences and the green water yard are located. The blending effect created by the landscape and waterscape used is made possible through optimized use of the water resource.

Natural treatment of walls embodies harmonious coexistence between residents and their natural setting while balcony lines provide rhythm. The rear elevation of the multi-layered structures emphasizes their massing, to reveal poetic interplay between structures.

北京昌平兰亭曲水流觞

BEIJING ZHONGGUANCUN TECHNOLOGY ZOO CHANGPING GARDEN MATCHING RESIDENCE B PROJECT-LANGTING~QUSHUILIUSHANG

该项目位于北京市昌平区，距八达岭高速公路南侧约600m，紧邻高速路出口，距未来地铁站1.5km。

本项目以传统方格网布局，以四合院作为空间构成的基本母题，形成适度的组团规模，并取《兰亭序》中"曲水流觞"和风水学中"曲水"之意，在地块中心形成带状景观区域。

建筑造型取意于"四水归堂"之说，坡顶向院内倾斜。设计中建筑造型采用传统的建筑元素：白墙、灰瓦、坡顶、棱窗，用简约的手法表达其意境，突出传统人文气氛。

建筑注重细部设计，体现"巧于因借，精在体宜"，同时吸收了南方园林"游廊"空间，在院墙外围设置了"游廊"，使院墙不再枯燥，增添了建筑光影变化，同时具有可游、可赏、可歇的功能。

建筑面积：99469.1m²
Total Floor Area：99469.1m²
设计时间：2005 年
Design Period：2005

This project is located in Changping district, approximately 600m south of the Badaling speedway, adjacent to its exit, and 1.5km away from the future subway station.

Laid out in a traditional grid pattern, the base concept of the courtyard house style is adopted to form the ideal group scale.

From "Sishuiguitang", the roof is inclined inward. Traditional construction elements such as white wall, grey tile, pitched roof and arris windows are adopted in the design, expressing an artistic concept in a simple style to enhance its traditional mood.

Careful attention to detail is observed to create the essence of "travelling corridors". The "travelling corridors" set on the exterior walls does not only amplify interplay of light and shadow, it also promotes dynamism and enhances and feelings of enjoyment and relaxation.

广州中海名都
ZHONGHAI MINGDU, GUANGZHOU

该项目位于广州市海珠区纺织路1号，北临珠江及40m宽沿江绿化带，东侧为江湾路，江湾大桥直抵用地东北角，地块西侧为30m宽规划路。总用地面积为92315m²，分四期开发。

项目充分利用基地周围的环境景观，引入"生态"概念，做好高密度控制下的人文居住环境。高层住宅采用了单体并联形式，有效地增加了使用面积，扩大了空间环境。在设计中以创造完美的居住环境和丰富城市景观为前提；以创造完整统一并独具特色及时代感的造型为目的；以合理的平面布局，美观耐用、技术先进的设备选型为原则；使本工程建成之后，在环境空间及造型上处于时代的前端。

建筑面积：401649m²
Total Floor Area：401649m²
设计时间：2001 — 2004 年
Design Period：2001 — 2004
竣工时间：2004 年
Completion Period：2004

合作设计单位：新加坡雅科本
In Cooperation with：Singapor Archurban

获奖情况 Awards：
2005 詹天佑金奖
Zhan Tian You Golden Award，2005

The site covers an area of 92315m² and is located along Fangzhi road in Haizu district, Guangzhou. To its north is Zhu river and a 40m wide greenbelt. To achieve an optimum living environment, the project makes full use of existing sights and complementary landscaping.

Parallel connections in high residences increase the utility areas and enlarge its surrounding spaces efficiently. A precondition to creating the ideal living environment for the site and enriching urban views around it is the objective of creating an image of a complete, unified, and unique built environment. The combination of a sensible site layout, classic and elegant elevation designs, and modern construction technology would catapult the project among the top residential communities in china.

南京中海·塞纳丽舍
CHINA OVERSEAS SENALYSHE GARDEN, NANJING

南京中海·塞纳丽舍位于南京市河西新城区的中部，南湖路与文体路之间。是中海集团在南京河西新区开发的一个大型中高层低密度的景观社区。总占地面积121668.6m²，建筑面积约为210000m²。

黄山路从小区中间穿过，把基地分成两个部分。在景观规划布局中，以景观为中心，以功能为先导，采用南偏东10°的布局方式。使其既符合南京人的居住朝向观念，又能满足日照通风的要求。中高层住宅的形态，既具有多层住宅亲切宜人、房型好、得房率高的特点，又综合了高层住宅结构强度高、耐用年限长的特点。一梯两户的平面布局，使得房间南北通透，通风、采光、景观效果都得到极大满足。

单体设计遵循经济适用、布局科学合理、使用方便舒适、功能齐全、美观高雅的设计原则，卧室、客厅尽可能朝南布置，南北通透。所有单元均采用一梯两户的布局。户型面积由85～220m²不等。部分景观好的单元采用复式设计。在整个用地的西北角考虑了社区中心的用地，在南部下角布置了幼儿园，占地4500m²。商业考虑为一楼的底商，沿五环路为主要商业面，黄山路靠五环路一侧也有部分商业延伸。

经典的古典三段式立面构成，加上对稳定、色彩、细部线脚及古典法式元素的合理利用，形成了新古典法式风情建筑立面，给人以稳重高雅的舒适感觉。

建筑面积：210000m²
Total Floor Area：210000m²
设计时间：2006年
Design Period：2006

The project site, with a floor space and covered area of 121668.6m² and 220000m² respectively is a wide, upscale, low-density community developed by the Zhonghai Group.

Huangshan road divides the base into two parts. The 10-degree south-by-east distribution uses the area's natural sights as its focal point, and functions as the planning guide to achieve a concept of a distinct Nanjing lifestyle. The construction is pleasant not only in its geometry, but it also serves it functions well. The plane distribution of one-step, two-residence provides each residence with optimum access to visual resource and natural ventilation as well as a delightful orientation.

Residential units vary from 85m² to 220m², with some units having two-level layouts. A kindergarten with an area of 4500m² is situated in the south. The main commercial area is distributed along the five-ring road, with some parts extending up to Huangshan road.

The new-classical French elevation is achieved by the incorporation of segmented vertical planes, and other French elements for an image of that timeless, upscale elegance.

东莞西湖春晓
WEST LAKE SPRING MORNING, DONGGUAN

该项目位于东莞新城市中心区，地处南四环路与东莞大道交会处，景湖花园以北，东莞大道西侧，交通便利。小区由11栋多层住宅，39栋中高层和5栋高层住宅组成。南侧设置小学和幼儿园，考虑以后可以为周边地区提供配套，小学规模为36个班，幼儿园为12个班。

项目造型追求创新，采用荷兰风格的现代建筑形式，通过整幅落地玻璃窗和凸窗角窗设计，使整个建筑通透而简洁，配合水平向白色阳台和挑板，以及顶层通高两层的玻璃面，形成了虚实对比，富有现代感的建筑风格。墙面采用与玻璃相近颜色面砖同黄色面砖的强烈对比，区分出质感与色感的变化，体现出柔和与硬朗的共生。

建筑面积：378014m²
Total Floor Area：378014m²
设计时间：2003 年
Design Period：2003

The project is in the centre of Dongguan with convenient traffic access. It consists of multi-level residences, schools, and other support facilities.

For this project, a Scandinavian theme is adopted - particularly that of the Netherlands. French glass windows make the construction bright and simple, matching attractively with its white balconies; the balcony floors and its glass surfaces exude a modern construction style. The strong contrast between glass colour and the yellow of bricks create texture that stimulates the senses and recalls its Scandinavian theme.

南京苏源颐和美地（南园）
NANJING SUYUAN YIHE MEIDI (SOUTH GARDEN)

该项目地处南京市江宁经济技术开发区，内环路以南，九龙湖路以东，诚信大道以北，殷巷路以西。

项目基地周边自然环境良好，其西侧及南侧皆有直接毗邻的大型城市景观公园及九龙湖自然景观。项目主要以独立屋、独立别墅以及部分高层、中高层住宅为主；规划构思明确建立"以水为悦"的营构主题；丰富的九龙湖水景资源是本项目基址依托的自然景观元素；临水而居是本社区的最大特色。因此，设计力图表达居住者对大自然美景的积极回应与主动对话，通过各组团之间丰富多姿的水景营造，充分展现人们对"水"的钟爱与依恋。

一派田园风光且又现代化的聚居理念，在相当长一段时间内都将是人们追求的目标。

建筑面积：418000m²
Total Floor Area: 418000m²

设计时间：2004 年
Design Period: 2004

This project is located in the Jiangning Economic Technical Development Area, defined by Inner-ring road on its south, Nine-dragon road on its east, Chengxin road on its north, and Yinxiang road on its west.

The site has an excellent natural environment; its west-south corridor looks out to a large urban park and the Nine-dragon Lake Park. The project fosters detached houses, villas, and some multi-level structures. "Enjoyment from water" is the general theme adopted for site planning. Abundant water source coming from the Nine-dragon Lake is the planning foundation and its position relative to the lake is the site's most salient feature. Therefore, planning naturally focuses on providing harmony between the residences and its well-endowed natural environment as well as optimizing the water feature resource by creating interesting waterscapes throughout the site.

南京星雨花都
XINGYU HUADU, NANJING

该项目位于南京市河西新城区江东南路与集庆西路交会点的西南角。设计力求创造中国一流高尚居住社区新形象，打造时尚、健康、生态型人居社区的划时代产品。

项目具有优美的小区内部环境，小区中心是中央景观绿化区，充分强调中心庭院的共享，使每户住宅达到景观朝向利用的最大化。同时充分考虑地块特点，整个小区采取封闭式统一智能化管理。会所、学校、幼儿园、open-air步行街，既与小区住宅部分紧密联系，又在管理上完全分开。充分利用了公建对外的商业价值，在使小区实现了人车分流、内外分流、商住分流的同时，也使整个小区提升到一个更高的层次。

建筑面积：400000m²
Total Floor Area：400000m²
设计时间：2003 — 2004 年
Design Period：2003 — 2004

The project is situated in a new district, on the western side of a river in Nanjing. The design aims to create an upscale residential community defined by a healthy living environment and modern amenities.

With central landscaping, the site is provided with an excellent centre yard shared among residents where each residence is afforded with an optimal view towards it. In consideration of the special landform, a closed, dedicated management system is adopted. Clubs, schools, kindergartens, and pedestrian streets connect with the residence closely and yet, are separated by the special management system. Full access to public facilities contributes significantly the convenience of its residents and to the value of the area.

南京东方天郡
ORIENT TIANJUN GARDEN, NANJING

该项目位于南京市仙林大学城中心区内两侧坐拥中心区，北临大学园，南、东、北三面被约2km外的绿化山体环绕，自然景观资源非常优越。

基地被仙霞路分为A、B两块地，项目定位为高品质时尚住区，在保证朝向、日照间距的前提下，规划中特别注意各户型的均好性与内外景观资源的充分利用。住宅布局遵循"南高北低、西高东低"的大趋势，使总体布局与大学城中心区的规划相协调，为城市空间添彩，且最大限度地利用周边的资源景观，创造出自身鲜明的群体形象。

根据用地的几何特点，兼顾与周边地区的关系，将住宅分为高层带、中高层带与低层多层区，三部分互相咬合，形成变化丰富的沿街立面与城市天际轮廓线。

建筑面积：410000m²
Total Floor Area：410000m²
设计时间：2004 年
Design Period：2004

The project site, located in Nanjing, is endowed with excellent natural sights. The site is divided into sections, A & B by Xianxia road. A modern residential district is the goal for this project. Optimum use of natural views is pursued while ensuring proper building orientation and adequate sunlight distribution for all residential units. Adopting the trend "sunken north and east, elevated south and west" matches the site plan with the campus plan and adds color and vigor to the cityscape.

The combination of residences, which are distributed according to the geometric characteristics of the landmass, sculpts a rich contour line for both the proposed project and the city.

苏州埃拉国际·自由水岸
AILA INTERNATIONAL LIBERTY BANK, SUZHOU

该项目位于苏州市郊金鸡湖东侧，工业园二区（一个包括居住、商业、学校等2万人口的综合居住区）的东北角。

项目用地分为A、B两区，A区采取高层低密度布置方式；B区中高层采用板、点组合体方式布置，通过两区不同的规划布局，使之产生不同类型的环境空间和意象。

苏州园林以"小桥流水"为主要特征，"临水而居"成为住宅布局的重要方式。本项目通过建立"水"的主题进一步发挥这一传统，同时提高小区的品质。

设计通过对比、衬托、借景以及尺度的变换、层次的配合，将苏州传统民居意境融入现代小区环境之中，并着重环境的立体化及生态性，绿化不只存在于地面，更向高空及地下延伸。住宅楼层除容纳基本居住活动之外，更有大量空中花园及入户花园，地下车库增设多处景观采光口，并配以种植、水面、室外楼梯，形成多层次、全方位绿化体系。环境小品精致、轻巧，使居民享受来自传统文化与工业文明的双重愉悦。

建筑面积：250000m²
Total Floor Area: 250000m²
设计时间：2003年
Design Period: 2003

This project sits east of Jinji Lake in Suzhou, northeast of the Industrial Zone. It has a residential district, a school, some commercial spaces, and an incumbent population of 20000.

The project is segregated into areas A & B. Area A is comprised of low-density structural distribution while B is composed of a mixture of individual structures and connected ones. This programming layout is intended to generate a different environmental design and a unique urban image.

Waterfront living is the main theme adopted for planning where the project builds and develops on the topic of "water" to improve scenery for the district. In addition, new variables such as scene sharing, scalar manipulation, a mixture of layers, and traditional elements distinct to Suzhou are applied into the site development to create an integrated and stimulating environment.

Elevation design and ecological awareness are design constants. Landscaping is done both on the surface and at sub-street levels. Light wells, planting strips are incorporated throughout the plan and even on stairwells to complete the planned ecological system of the community.

广州光大花园
EVERBRIGHT GARDEN, GUANGZHOU

该项目位于广州市海珠区，规划中充分利用基地内已有高大榕树林及丰富的地形高差，将小区设计成一个环境优雅、空间宜人的"大榕树下健康人家"。

总图上革新路与另两条规划路将用地分为四块，自然形成四个分区。通过两条带形绿化公园将四个分区连接，建立社区公共空间的基本构架，形成完整的户外活动空间"链"，将居住区保留的生态绿化、人工绿化以及珠江风景联成一个整体。

景观视觉走廊：构想了由东到西至江边，由南到北至古炮台公园，由入口迎宾道至社区礼仪文化广场至中心会所至江边的三条视线走廊，并将高层住宅与此相应布置。

城市设计：着重考虑高层住宅的整体布局方式以形成具有视觉趣味的天际线和视线走廊，并以流动的形式引导人流，组合主要公共空间。景观上注重形成开阔的视野以及造景和借景。

建筑面积：1000000m²
Total Floor Area：1000000m²
设计时间：1998－2003年
Design Period：1998－2003
竣工时间：2003年
Completion Period：2003

This project is situated in Haizhu district, Guangzhou. Proposed as "Healthy residences under the banyan", it sits on a lush natural setting highly suitable for residential use which is further enhanced by existing verdant banyan trees and stimulating natural landforms.

Gexin road and two other minor roads divide the base into four natural blocks. Twin green belts connect the four districts to build the basic frame of a communal urban space. This creates a link between public human activities, their natural setting and the project development.

Visual corridors are created to complement the high level residences, namely: east-west to the riverbanks, south-north to the ancient park, and finally, a linear corridor from Yingbin road entrance, to the cultural plaza, all the way out the riverbank on the far side.

Urban design invests substantial consideration on the overall layout of the high-rise residences to enhance its skyline and visual corridor. It also renders emphasis on the integration of core public spaces by providing fluent circulation to and within the site. Landscape design strives for wide open views, as well harmonious interaction with surrounding architecture.

成都花样年·花郡
CHENGDU HUAYANGNIAN FLOWERY COUNTY

本项目定位为"满足居家需求的大众精品住宅,东二环30万m²的大社区"。设计指导思想是充分利用基地内在的环境景观引入"生态"概念,做好高密度控制下的人文居住环境,构建和谐社区。

设计特点:总平面布局在满足容积率及建筑密度的条件下创造出怡人的居住空间。经过多方案的论证,在满足规划条件及业主具体要求前提下,采用合理用地的周边式布局,使小区由4个周边式组团串联而成,形成相对完整独立,有利分期开发的规划形态,商业裙房沿周边展开。住宅布置尽量做到景观的均好性,最大限度地利用内向的庭院景观,避免楼与楼之间的对视干扰,布局合理,通风采光符合成都当地生活习惯。

每期住宅开发,围合形成一个中心花园空间,满足人与自然的亲近,使社区从繁华、喧闹的都市中分离,形成一个独特的生活空间,营造宁静、祥和的环境。小区环境基于开放型的空间设计理念,强调生态、注重环保。绿地环境相对集中,部分住宅首层局部架空绿化,疏透开朗的空间与充满生机的自然环境结合。

立面风格追求自然,摒弃多余的装饰,回归建筑本质,反映居住与自然良好关系。主墙面为浅灰色和白色,错落和凹凸的墙面为少量跳色,对比中保持协调,有层次感而不失整体、轻重搭配、明快自然。

建筑面积:358300m²
Total Floor Area: 358300m²

设计时间:2006年
Design Period: 2006

The project is defined as "The 300000m² upscale residential district of the east." Making full use of the environment and importing ecology to create a soulful residence within a harmonious community is the main guideline used for project design.

The whole layout creates a pleasant residential community while it satisfies structural requirements on construction density and FAR. Its "circumjacent" layout makes the district both isolated and integrated, and ideal for a staggered development program. Site layout pursues unobstructed views and ample lighting and aeration.

A central garden which is formed and enlarged by a series of enclosures from each stage of development makes the community both private and quiet. Its landscaping stresses on environmental protection and ecological responsibility. The combination of natural flora and planned landscaping creates a pleasant space for a residential community.

Natural treatment of its walls reflects the good relationship between built structures and nature. The visual expression created from light-grey and white walls makes the construction seem brighter and more natural.

南昌金域名都
JINYU MINGDU NANCHANG

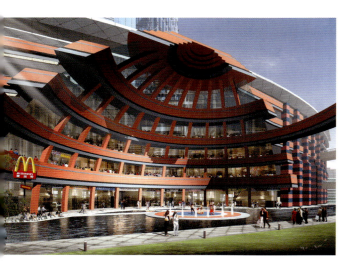

该项目位于南昌市洪都大道以东,北京东路以北(原南昌市体育场用地),北临建设中的体育公园,西邻省图书馆,东面为城市滨水体——玉带河所围绕。周围配套设施齐备,是理想的居住、办公及商业用地。

项目突出"现代、闲适、文化、优雅"的格调,并进行拓展和延伸,大胆引入澳大利亚黄金海岸"浪漫、清雅"的地域环境理念及人居特色,体现"浪漫、经典、时尚、个性"的住区风格。

湖景位于基地的北端,已建成的玉带河位于基地的东侧。设计以一种较大的手笔对两者进行有机的形态联系,最终确立了以曲线为主,南北两组空间渗透、流动、点板结合的空间走向。住区的北端酒店式公寓、南端纯写字楼都以板式高层作为空间序列的起止,此布局基于两个端头的景观与其他位置不同,打破空间布局的均质化,体现其景观价值的差异化特征。

建筑面积:368800m²
Total Floor Area: 368800m²
设计时间:2003 — 2004 年
Design Period: 2003 — 2004

Situated in Nanchang, to its north is a Fitness Park, west is a provincial library, and east is the Yudai River. The project site has complete access to public and support facilities which makes it a convenient place for residential development.

Lifestyle traits on fashion, leisure, culture combined with environmental concepts on romance and elegance creates a whole new dimension of urban living.

A lake located north of the site and the Yudai River is in the east. In order to connect these two resources, a curved, open space was established in the north-south quadrant. The site plan which plotted the different positions of the scenic spots at the end nodes breaks the monotony of the space while showing the different specialties of the area.

大连铁龙·动力院景
DALIAN TIELONG DONG LI YUAN JING

本项目位于大连城市发展主城区和新城区的连接区域——甘井子区，地处城市两条交通主干道华北路和华东路之间，是大连市"西扩北进"战略中在北部华南板块重点开发的大型住宅小区。

方案充分利用现状中优良的自然景源，从中提炼出山、水、庭院等元素作为构成整个楼盘构架的脉络，使整个楼盘成为一个有机的整体。在整体和细节中注入健康及生态的元素，打造一个将自然生态和城市文化融为一体的完美居住区。

建筑风格简洁庄重，空间富于特色，住宅建筑设计均考虑良好的景观朝向，视野开阔、环境优美。同时，运用对景、借景形成空间层次。采用"社区半开放、组团封闭"的社区管理理念，公共空间与院落空间相对独立。

建筑面积：158000m²
Total Floor Area：158000m²
设计时间：2006 年
Design Period：2006

The project is located in Ganjingzi District which is at the junction of the main district and the new district of Dalian City. Its unique location and convenient access to two main city streets (Huabei and Huadong Avenue) determines a new standard for large-scale residential areas in the city.

Imbued with the elements of 'hill, water, and courtyard', its natural setting provides essential resources to build up an organic community. The idea of 'health and ecology' is imbibed into the whole project where each detail will contribute to a pleasant living space filled with modern amenities and cultural features.

Proper orientation and access to natural scenery are the key points considered for the apartment layouts. The facade is simple and elegant with diverse spatial treatments. The community will be managed based on the concept of a semi-open community with an enclosed organization so as to make the public and courtyard spaces effectively segregated.

南京麒麟山庄
NANJING KYLIN VILLAGE

南京麒麟山庄东郊花城地处南京市江宁区麒麟镇，宁杭公路以南，青西村以北，麒麟门村以西。

规划基本上分为三个高度及层次：多层及高层主要布置于基地之北面及最南面，亦是基地之高点，所有单位之景观主要是南向，从高处能欣赏湖区景色，湖畔别墅景色尽收眼底。小高层与多层之间是宽阔之绿化小区，园林布置精巧，也是居民休息、玩耍之场所。从多层至湖景别墅区之间，是低密度的低层，因地形级级向湖，从而都能享受到湖景及园林景色。小区绿化的组成主要由别墅及连排别墅的私家花园、组团绿地、中央景观及湖畔长廊、房前屋后绿地、半地下室车库屋顶花园、顶层露台及屋顶花园，共同组成了完备的绿化生态系统。

规划以生态优先、整体优先为原则，注重考虑环境保护、生态平衡，使人与大自然有机结合，充分利用地势及自然环境，创造湖水区，利于雨水储备作为绿化之用。建筑方面，住宅每户主要朝向为南向，尽量能带动对流作用，连排别墅更设有玻璃光顶，可促成冬暖夏凉之效果，尽量设置太阳能热水器。小高层底层采用框架结构架空处理，使空气流通更顺畅。

建筑面积：465364m²
Total Floor Area: 465364m²
设计时间：2005年
Design Period: 2005

The project is located in the town Kylin, Jiangning district in Nanjing.

The project requires planning and design for multiple typologies: a) high- rise residential structures situated in the north-south quadrant of the site. These structures are oriented towards the south where an excellent view of the lake is accessible, especially to the upper-level units. b) a landscaped area between the high-rise structures, ideal for outdoor activities, relaxation and social interaction. Private gardens, roof gardens, pocket parks, and a central garden complete the landscape system. c) low-density, low-rise residential structures are situated between the lake and the central garden so these residences are able to enjoy both visual amenities from the garden and the lake. Prioritizing the ecology and the community achieves harmonious co-existence between man and nature. The lake obviously provides distinct advantages to site development and to landscaping so site layout makes optimum use of this natural feature by setting building orientation towards it where light and air ventilation are also optimal.

重庆华宇·渝州新都
HUAYU YUZHOU XINDU, CHONGQING

　　该项目位于渝中区大石路原普天通讯设备有限公司厂址，占地面积约74854m²。

　　小区沿大石路布置商业街、核心商场，充分发挥城市主干道的商业价值。基地西南侧，在大石路和规划路的交会处，布置一幢30层高的公寓商住楼，其显赫的地理位置及新颖的造型使其成为该小区的标志性建筑。与其相对应的基地东南面，则是一幢与商业街连为一体的弧形公寓楼，该建筑优美而舒展的造型，与西边的公寓楼相呼应，形成别具一格的城市景观。

　　住宅的布局适应重庆地区居住建筑的特点，以满足住户最好的日照条件和力争最好的景观为出发点，采用"折线式"和"点板结合"布局，在顺应复杂多变高低错落的用地现状的同时，又有机围合出一个中心大花园，两个组团绿化庭院，形成"一大两小"的环境模式。从建筑肌理上让小区形象给人以耳目一新的感觉。

建筑面积：312000m²
Total Floor Area：312000m²
设计时间：2003—2004年
Design Period：2003—2004

The project covers an area of 74854m². It is situated within the compound of Putian Communication Equipment Co., Ltd on Dashi road in Yuzhong district.

The Commercial Street and a core market are set along Dashi road to enhance the commercial value of the main city road. A thirty floor commercial-residential building is set at the crossing of Dashi road and Programming road and a curved apartment building is located in the southeast section of the site. The spatial relation shaped by these two buildings creates a new landmark for the city.

The Site layout and building orientation is adapted to the geographical characteristics of Chongqing using sunlight and scenery as key points for planning. "Folded style" and "spot-board distribution" is also used to adapt to the changing contour lines and create the mould for its core space. Its central garden and two pocket parks complete the environmental concept that offers a refreshing atmosphere to its residents.

南宁佳得鑫·水晶城
JIADEXIN PLAZA, NANNING

该项目位于南宁市新市中心区，为集高档主题休闲、娱乐、商城、地下大型购物中心、城市中心高档住宅小区于一体的大型综合建筑。

设计结合地方特点，充分考虑户内通风与采光，明厨明厕，户型平面方正实用，分区合理。户内设面积较大的花园式生活阳台及宽敞实用的服务阳台，采用低窗台凸窗，宽敞明亮。

简洁而生动的新风格，富于动感的建筑形体，以现代建筑气质，突出体量特征。通过阳台和建筑上部的特殊处理形成丰富有致的建筑轮廓线，建筑线条强调水平与垂直的均衡。

佳得鑫广场用地似弯曲的梯形，北边长、南边短。本项目着重城市文脉关系，将17万m²的巨大建筑体量布置在基地北面，约占基地三分之二；基地东南三分之一为下沉式休闲广场，以此和城市亲切对话。

建筑面积：170000m²
Total Floor Area: 170000m²
设计时间：2002–2003 年
Design Period: 2002–2003
竣工时间：2005 年
Completion Period: 2005

Situated in the new centre of Nanning, it is a large integrated construction that contains residences, leisure and entertainment facilities, a mall, and an underground shopping centre.

Combined with indigenous elements and full consideration of light and air ventilation, these ideal residences make its isolation reasonable. The garden style balcony, the utility area and the low bay windows creates a bright and airy atmosphere for the residences.

The shape of the structure and the vigour of modern architecture accentuate the structural massing. Contour lines are formed from veranda details where construction lines stress the equilibrium between the horizontal and the vertical.

Jiadexin Square's base is shaped like a wedge where its northern side is longer than that of its southern side. Site layout allocates two-thirds of the total area to structural development while one-third is dedicated to open spaces like piazzas and gardens where social interaction takes place.

深圳中海宝安松岗
CHINA OVERSEAS BAO' AN SONGGANG, SHENZHEN

项目用地位于深圳宝安区松岗街道，东临郎碧路，南靠（新）沙江路和河滨路，东南向有松岗街道办、中心广场、文化广场、佳华商场等，北邻（老）沙江路，西部隔70m绿化带邻广深高速公路。

以平行于广深高速公路的东南向主轴作为基准，使住宅建筑主要朝向均避开不利方位，迎向景观、日照、通风的最优方位，从而取得视觉上、生理上、心理上的三重优化。体量组合方式采用西低东高、南低北高以及西、北密闭，东、南开放的模式。以错动为基本模式，以序列化的组织方式形成三次元空间利用的景观布局模式，在有序的基础上实现空间的流动和主题的更替。

户型分布建立在居住用地价值评估的基础上，在体量正向分布的合理性基础上再次叠加户型正向分布配置，对基地的土地价值潜力作更进一步的挖掘，同时也强调均好性的重要性。

The project sits along Songgang Street in Bao'an district, Shenzhen. Land mass is elevated at the north-east sector and sunken at the south-west.

Using southeast orientation as the norm, the development will have excellent views, adequate sunshine, and good ventilation. Its excellent site features optimizes the area's image, physiology, and psychology.

Distribution of residences is built on a careful evaluation of land characteristics to highlight planning principles on harmony between the built environment and its natural setting.

建筑面积：200000m²
Total Floor Area：200000m²
设计时间：2005 年
Design Period：2005

东莞凯达华庭
KAIDA HUA TING, DONGGUAN

　　东莞凯达华庭位于莲湖之滨、松山之畔，整个莲湖面积浩大，种满婀娜多姿的莲荷，夏季是整个桥头乃至周边地区一个上佳的度假休闲之地。松山公园一年四季绿荫葱葱，也是当地的一景，山水交融，成为本项目的一个极好的外在景观因素。

　　设计通过灵活多样的设计手法努力营造一个丰富多彩的建筑体系和景观体系，小区设计与外在的景观建立一种内在的联系，创造出一个既有住家温馨又有度假休闲浪漫气息的现代生态园林社区。

　　整个小区采用南低北高的建筑布局，南面紧邻莲湖，布置别墅区，使别墅区有"面水靠山"的感觉。别墅区本身也采用南低北高、突出中心的布局。前排临湖别墅采用两层建筑，周边及中心大别墅采用3层建筑，使莲湖景观尽量为所有别墅所享受。

　　结合临湖景观，临街设置部分高档商业铺面，商铺的天台用作公共观景平台，使小区与景观街、莲湖产生更为紧密的联系。

建筑面积：200000m²
Total Floor Area：200000m²
设计时间：2002年
Design Period：2002
竣工时间：2006年
Completion Period：2006

The project is situated on the banks of Lian Lake and Song Mountain. The Lian Lake which is famous for its delightful lotus verdure is an excellent holiday spot for surrounding communities. Song Mountain which is green and lush each season is also a scene in itself. Taken together, these natural sceneries create a fascinating sight

This natural spectacle, combined with scenery from within the project site (landscaping) creates one truly ecological community.

The construction layout of "low in south and high in north" is adopted. Setting the villas in south where it abuts Lian Lake creates a feeling of "Leaning against the mountain and facing the water". The villa layout also adapted adopted the tact of extruding the centre part. This way, the two-level villas are situated in the front row while three-level villas are located in the centre so surrounding views can be enjoyed by all residential units.

Top grade commercial shops are located with visual access to the lake, where shop verandas may also be used for sightseeing. This situation takes the community even closer to the lake and entices people to visit the shops located within the area.

武汉东湖·香榭水岸
WUHAN EAST LAKE WATER FRONT

该项目地处湖北省直机关所在地——水果湖地区，滨临风景优美，高校环绕的东湖，紧靠迎宾大道。地势平坦，东西向长约180m，南北向长约136m，规划用地约21808.82m^2。

小区由三栋住宅组合成半围合状，形成安静温馨的内聚空间——入口广场和中央花园，灵活通透。建筑尽量外扩布置，取得最大的庭院空间。建筑底层大部分架空，成为休闲绿化场所，增大中央花园的进深感，并且能使夏季主导风向引入内庭，形成良好的空气对流。

小区的户型设计充分利用了基地的日照方向与景观方向的一致性，使多户在拥有良好的日照、通风条件的同时，充分考虑了对景观的需求。

建筑群体造型力求突出简洁、鲜明，并带有清新、飘逸的滨水建筑特色。建筑的连体布置形成有节奏的韵律，并通过高低错落的变化，形成富有活力的建筑群，并展现优美的群体轮廓线。

建筑面积：85514m^2
Total Floor Area: 85514m^2
设计时间：2002年
Design Period: 2002

The construction whose planning area is 21808.82m^2 located in the seat of provincial government-Shuiguohu District, with a beautiful landscape and flat landform.

The quiet and warm inner space-entrance square and centre garden was formed by the semi-enclosure of three residence buildings. The exterior-enlarged style construction got more space. The bottom of the construction which built in stilt became the virescence space and formed the good cross-ventilation.

Making good use of consideration of consistency of sunlight and sight, the design ensures good sunlight, aeration and sight.

The figure of the construction pursued simple, bright and the features of waterside. The continuous groups which formed by the layout of the construction and the change of the different height showed the excellent colony contour line.

南京汇林绿洲二期
PHASE II OF NANJING HUILIN OASIS

该项目位于南京城市次中心地带，北临黑龙江路，东临芦席营路。

本项目采用"大花园"的规划原则，使水景、绿景向小区内部延伸，呈互补关联，力求"水景利用最大化"和"小区整体内外环境景观的均好性"，使"水韵、绿意"无处不在。

利用不同建筑高度的日照间距，使左右两列小高层呈错列式布局，以便整体空间错落有致，形态活跃而不呆板。

采用现代风格的设计手法，突出线条与地面的穿插，线与面两种设计元素的对比，既丰富了建筑物的空间形态，又体现了建筑物的抽象感、时代感、生态感。商业用房采用虚实对比的手法，透明生动的玻璃幕墙与金属板挑檐相互对应，展现出建筑轻盈通透的特征。

塔楼部分用凸窗、转角窗、落地窗的设计手法，运用竖向和横向板带线条将阳台凸窗连成一体，屋顶构架设计明快、简洁，极大地丰富了塔楼顶部的空间层次。主墙面为浅暖灰色调，创造了平和亲近的立面风格。

建筑面积：125286m²
Total Floor Area：125286m²
设计时间：2003 — 2004 年
Design Period：2003 — 2004
竣工时间：2005 年
Completion Period：2005

Situated in Nanjing, to its north is Heilongjiang road, and to its east is Luxiying road.

Adopting a "Large garden" layout principle, the site plan brings waterscape and landscape into the site to interact with each other. Pursuing optimum use of the water feature and the whole landscaping makes these elements largely accessible and available throughout the site.

Varying heights and penetrating sunlight promotes an appearance of interlaced structures and thus, creates a dynamic and lively structural form.

Elevation detail extrudes lines and surfaces. This treatment of lines and surfaces does not only enrich the space, it also enhances a feeling of abstraction and style. The introduction of bay windows, corner windows, and the simplicity and brightness of the roof construction adds depth to the spaces at the upper levels.

深圳中信红树湾花城北地块
CITIC MANGROVE BAY FLOWER CITY NORTH PLOT, SHENZHEN

该项目位于深圳南山区，沙河东路与白石路交界处。用地面积80.7万m²，总建筑面积32.1万m²。本工程包括7幢高层住宅、1幢高层公寓及3幢联排别墅、幼儿园、会所、地面绿化广场、地下停车库及设备用房。共提供1385个住宅单位，户型面积为40～370m²。

地块西临环境优美的沙河高尔夫球场，东面为新建的居住及商业文化设施，南可远眺深圳湾海景，景观资源十分丰厚。

建筑布局均为东南、西南走向及正南朝向，绝大多数住宅拥有高尔夫景及海景。使建筑群充满韵律感、节奏感及视觉层次感。

入户花园、内花园、观景阳台成为户型设计的亮点，建筑采用借景与造景的手法将室外绿化有机地组织于住宅中。采用较大的落地玻璃窗，赋予充足的透明度和轻快的时代感，建筑色彩以深灰色配以白色明快线条，淡雅清新，富有时代气息，为整个小区平添浪漫色彩。

建筑面积：321000m²
Total Floor Area：321000m²
设计时间：2005年
Design Period：2005

合作设计单位：
澳大利亚柏涛（悉尼）建筑师事务所
In Cooperation with:
Peddle Thorpe Design Office

The project, whose floor space and covered area are 807000m² and 321000m² respectively, is situated at the junction of Shahe east road and Baishi road. It contains 7 high-rise residences, 1 high-layer flat and 3 townhouses, a kindergarten, clubs, landscaping, an underground carport and storage facilities. There are 1385 residences with areas varying from 40 to 370m² on offer in this project. West of the site is a golf course, to its east is a new residential community with commercial and cultural facilities, and to its south is an ocean view.

Building orientation towards the southeast, the southwest and south offers most residences with a sight of the golf course and the ocean, and it allows for a practical grid-like site pattern.

Indoor gardens and sightseeing terraces are key design features. Outdoor landscaping is organised into the residences by the tact of shared courtyards. Large French windows make the residences bright and modern and the combination of a white and grey palette paints a modern community of quiet elegance.

深圳京基御景华城
SHENZHEN JINGJI YUJING HUACHENG

该项目位于深圳市福田区繁华地带，华强南路和滨河路交会处，占地26466m²。该地块位于滨河路南边，紧邻香港，视野开阔，沿深圳河有极好的自然景观，且沿街具有极好的商业开发价值。

本项目在立面设计上试图摆脱中国住宅的可识别性和固定模式，强调建筑的简洁、屋顶和外墙的衔接统一、开窗的韵律感等。并在局部（形象入口处）采用新旧对比的手法加强标志感，强调立体绿化带来的视觉上的生态效果。建筑色彩特别是外墙色彩采用深色基调，并通过深浅对比丰富建筑的外部形态。

建筑面积：260000m²
Total Floor Area：260000m²
设计时间：2003 — 2004 年
Design Period：2003 — 2004
竣工时间：2005 年
Completion Period：2005

The project covers an area of 26466m². It is situated within a thriving community in Fu Tian district, at the crossing of Huaquiang and Binhe roads, adjacent to Hong Kong. It has wide, excellent, natural views along Shenzhen River and a promising commercial development value.

The project seeks to break out of the typical design mould and aims to set a new standard in Chinese residential design, emphasizing on simplicity, the unification of the roofing element to its body, and the rhythm of the windows. Contrast between new and old enhances the feeling of well being and stresses ecological responsibility. Warm coloured walls and its reaction to light enriches the external appearance of the structures.

青岛海信慧园二期
HISENSE HUI GARDEN PHASE II, QINGDAO

该项目位于青岛市延安三路东侧，江西路南侧。基地南面为海信慧园一期住宅小区，东面为现状空地。

小区住宅由九栋高度不同的住宅呈半围合状，围合成近9000m²的生态花园。建筑的连体布置形成有节奏的韵律，并通过高低错落的变化，形成富有活力的建筑群并展现优美的群体轮廓线。中高层住宅采用弧形的连体设计，把一部分空间退让给城市。高层住宅则通过错位、后退、架空等手法来减少北面对江西路的压迫感。多样的住宅平面组合，让每户都拥有良好的朝向与景观。

建筑群体造型突出简洁、鲜明的海滨建筑特点，并适应青岛特有的文化氛围。立面以浅蓝色大面积玻璃、低窗台凸窗、白色的阳台栏杆、水平檐板为元素，通过半圆形及波浪形的界面交织成一幅优美的图画。建筑群体呈阶梯状的变化，形成丰富而具标志性的群体轮廓线，体现自然的"山与海的对话"。

建筑面积：108000m²
Total Floor Area：108000m²
设计时间：2001 — 2002 年
Design Period：2001 — 2002
竣工时间：2004 年
Completion Period：2004

The site is situated on the east of Yan'an Third road, south of Jiangxi Rd in Qingdao. South of the site is phase II of Hisense Hui Garden while a vacant lot lies to its east.

Nine residential buildings of varying heights enclose a semi-circle to form a 9000m² ecological garden. The collocation of structural construction composes the cadent rhythm, and its structures create an interesting contour through diversification of its heights. The arc design on the upper levels creates space for interaction and other urban activities. The tact of building on stilts is adopted to reduce the oppressing sensation from Jiangxi road. Multiple-level residences create ample opportunities for pleasant views and adequate sunlight for each residence.

Structural construction used is simple and bright to suitably adapt to the cultural atmosphere of Qingdao. Component areas using light blue glass, low-sill bay windows, white balcony balusters, and curtain plates creates unique and interesting interplay on the elevations. The scalar construction shapes and contour lines reveal the natural dialogue between "mountain and ocean".

上海新江湾 C1 地块
SHANGHAI XINJIANG BAY C1 PLOT

该项目位于上海原有的工业城区杨浦区，是"中央智力区"（杨浦区）的重要组成部分，占地面积188500m²。

该项目主干系统：由社区人工生态公园——"新江湾"文化广场和线性商业空间"Pocket-Park"组成的社区主干景观体系是开放人群活动的主要区域，生物多样性、文化多样性、活动多样性有机地组合成完整的开放性外部公共空间体系。分支系统：礼仪绿化广场、组团庭院、组团间小公园、花街等参与性绿化景观系统，构成社区居民的半公共外部空间体系；各种屋顶花园、公共平台花园构成社区的半私密外部空间体系；花园阳台、户内花园、阳光温室构成了社区的私密外部空间体系。

建筑面积：282765m²
Total Floor Area：282765m²
设计时间：2005 年
Design Period：2005

The project with a construction area of 188500m², is located in a distinguished section of the Yangpu district, Shanghai.

The project is conceptualized in hierarchies through a "trunk system" which incorporates the community with a zoological park, cultural piazzas, a linear commercial strip, and pocket parks intended as venues for communal activities. These diverse areas for biology, culture, and community collectively form an integrated public space that is the core of the community.

Branching out of the "trunk" are a series accessorial layers. Its top is comprised open squares, yards, parks, and flower streets. These areas form the second tier of community space classified as semi-public.

Roof gardens, balconies, indoor gardens, greenhouses, etc. comprise the bottom layer and rounds up the hierarchy of spaces.

长春威尼斯花园
VENICE GARDEN, CHANGCHUN

该住宅滨临环境优美的南湖，基地内有一城市干道"卫光街"穿越其中，将用地分为东西两区。其中一期已有87000m²建成，二期实际用地为15.77hm²。

本规划通过三条轴线将整个小区联为一体。蓝色的水轴平均宽度为15～20m，用水把居住区中心与湖滨相联系，充分利用了狭长又难用的湖滨地带；绿色的森林轴平均宽度为20～30m，该轴横贯东西，联系整个用地的各边角用地，使被城市道路分割出去的地块成为居住区有机的一部分。这两条轴线的交织将各邻里组团庭院联为一体，从而营造出集南湖、水面、绿色生态公园于一体、邻里共呼吸的大园林生态系统，成为独具特色的品牌特征。中轴线的设置是为了切中威尼斯的人文主题，通过柱廊、广场、购物步行街、步行街、双尖塔、方尖塔到滨湖的文化活动广场一气呵成，通过空间的起伏与开合变化，强调了以人为本的中心思想，使人文特色与生态特色交相辉映、贯穿始终。

建筑设计在简洁现代的基础上，加入了线条及山花屋顶等古典元素，营造出"威尼斯"的文化情调。规划特别考虑了建筑轮廓对南湖的影响，沿湖布置小别墅等低层建筑，向北逐次升高，从而形成非常有层次的湖岸景观。

建筑面积：238030m²
Total Floor Area: 238030m²
设计时间：1999－2000年
Design Period: 1999－2000
竣工时间：2003年
Completion Period: 2003

获奖情况 Awards:
2001年度长春市优秀规划设计一等奖
First prize of Excellent Programming design of Changchun, 2001

The construction including completed 870000m² in NO.1 and 15.77 practical area in NO.2 neighboured on South Lake, was divided into east and west areas by "Weiguang Street".

Three axes connect the districts as a whole. The intertexture of 15～20m water axes which connect the residences and lakeside and 20～30m forest axes which traverses east and west made the groups as a whole to create a zoology system combines south lake, water, and Zoology Park, and formed the brand speciality. Middle axes was set to hit the topic of humanism, the non-stop completion of colonnade, plaza, shopping Street, Pedestrian Street, double minaret, square minaret and the cultural plaza, the gurgitation and change of the space stress the clou of "centred human", making humanism and zoology specialities interlude.

The project creates the "Venice" cultural sentiment by the combination of simple and modern design and the elements of lines and pediment roof etc. Considering the effect to South Lake, low constructions such as little villas were set along the lake rising to north to form the lakeside sights.

北京华美橡树岭
HUAMEI OAK RESIDENTIAL DISTRICT, BEIJING

该项目位于北京市朝阳区双纬路和双经东路交叉口的西北角，设计中本着打造'精品建筑'的理念，进行精心设计，使其成为该地段的地标式建筑。

平面设计以提升居住品质为导向，务实创新，注重均好性。户型多样，每户都赠送大面积落地凸窗、阳光房，首层赠送地下私家花园，顶层赠送屋顶花园，提高了户型的实用率及卖点。

立面设计以创新为本，以耳目一新的建筑形象冲击客户的视觉，以个性化塑造产品的个性，关注细节，追求精致。住宅立面采用圆弧屋顶、大露台、大面积玻璃凸窗、阳光房的变化，形成现代的丰富的立面效果，玻璃映衬着绿树、流水，使整个小区浑然一体。

总建筑面积：43285m²
Total Floor Area：43285m²
设计时间：2005年
Design Period：2005

The project which became the symbol by elaborate design located in Chaoyang Beijing.

The design which is innovative and practical took advancing the residence quality as the guide, stressed all good. The type of the residence is multiple, large area of French windows, sunshine houses, private gardens and roof gardens would be given to the residence, and improved the utility and features.

Taking innovation as the principle in the vertical plane design, impacting clients' vision by the new figure of the construction, the construction pursued the perfect. The vertical plane effect formed by arc roof, large terrace, and large area bay windows and sunshine houses and the tress water formed a whole of the community.

深圳田园居别墅
TIANYUAN VILLA, SHENZHEN

该项目地处景色秀丽的银湖之畔，是一处高档的别墅住宅区。

由于基地为高差较大、地形较复杂的山地，本小区在总体布局上以充分利用土地为原则，尽可能地利用地形地貌，保护了自然，简洁的道路系统。小区出入口设在基地南向，道路顺地势环形布置，将小区分成几个别墅组团。在小区最低处的入口部分布置了会所、管理用房和技术用房，该地段的地貌将网球场、游泳池、中心花园组成区内优美的景观。整个小区别墅以银湖水景为依托，为群山环抱，南对湖水，建筑层层跌落，富有山区特有的风韵。

根据各自的地貌特征及细微变化，不同点上的别墅在剖面和平面设计上作了灵活调整，室内外空间非常丰富。玻璃通透、明净，再加上不规则石材的贴面，建筑现代而朴实，环境幽雅而宁静。红色的瓦屋面和浅色墙面的色彩主调衬托在绿树、青山、碧水之中，别具一格，景色绝佳。

建筑面积：20224m²
Total Floor Area：20224m²
设计时间：1995 — 1997 年
Design Period：1995 — 1997
竣工时间：1998 年
Completion Period：1998

获奖情况 Awards：
2000 年深圳市第九届优秀工程设计一等奖
2001 年广东省第十次优秀工程设计二等奖
2001 年度建设部部级优秀住宅和住宅小区设计二等奖
First prize, 9th "Excellent Building Design Award" for Shenzhen, 2000
Second prize, 10th "Excellent Building Design Award" for Guangdong, 2001
Second prize, "Excellent Residences and Designs for a Residential District" from the Ministry of Construction, 2001

The project, which is on the bank of Silver Lake, is a top-grade villa residential district. Due to severe slopes and the complexity of the landform, maximized land use is the principle adopted for site development. Sensible use of existing contours avoids invasive site development and dictated the general layout for site planning. The entrance and exit are set in the southern side of the plot. Roadways are laid-out circular (contour like) where it divides the district into several groupings. The club and administrative offices are located in the lower part of the site. It is also the site contours which organize the locations of the tennis courts, swimming pool, and the central garden as part of the area's scenic attractions.

Clear glass and stone treatments sculpts a structure of modern, simple elegance. Changing contour lines and varying architectural features create a rich spatial palette.

深圳银谷别墅
YINGU VILLA, SHENZHEN

该项目位于银湖旅游中心的一个山谷内,三面环山,坐北向南,西面的山谷口视野开阔,视线一直可延伸至城市中心区,环境十分优美。该中心主体建筑采用多个体型组合形式以减弱体量较大的感觉,公共空间富有变化;正面有山泉汇集而成的水面,形成独具特色的室外环境。

多层接待用房室与会议中心配套的集休闲、娱乐、住宿于一体的综合性建筑群。根据地形分为三块,以庭院围合、高低错落、穿插渗透的手法,使建筑群以较小的体量分散于山林之中。采用岭南建筑风格所特有的语言,与银湖现有的建筑风格相协调。独立式接待用房共分7种类型,适应了多种不同需要。其中F、G型为引进加拿大赛纳西公司轻型结构装配式别墅,别具异国情调。

本设计在国际会议中心与接待用房之间没有绿化地带分隔,配合道路两旁及接待用房庭院的绿地构成一个点、线、面结合完整的绿化系统,绿地率达到65%。

建筑面积:60300m²
Total Floor Area: 60300m²
设计时间:2000年
Design Period: 2000
竣工时间:2002年
Completion Period: 2002

Situated in a valley of Silver Lake travelling centre, enclosed by mountains in three sides, seated in north and face south, the sights in the west hatch is wide and can be extended to the centre of the city, with an excellent environment. Multi-shaped constructions debase the feeling of large mass, the public space is changeable, the water focused by spring formed special exterior environment.

It is the constructions combined multilayer reception rooms, meeting centre, leisure, entertainment and residences. Being divided into three blocks according to the landform, the ploys of yard enclosure, interlude and infiltration made the construction distributed in the mountain by small mass. The language of constructions in Lingnan made the project harmony with the existing style of Silver Lake. 7 different types of reception rooms satisfied different requirement. The style of F and G equipped with the light device of Canada show the foreign taste.

A complete virescence system was formed by green belts and the green plot in the sides of the roads. The virescence ratio is as much as 65%.

上海慧芝湖花园
HUIZHIHU GARDEN, SHANGHAI

本项目位于上海市闸北区共和新路以东，北宝兴路以西，广中路以北，规划灵石路以南，与著名学府上海大学仅一街之隔，该住宅区总用地149614.2m²，规划中平型关路将地块划为东西两块，其中东地块总用地为108300m²，容积率为2.5，建筑高度控制100m以下，用地平坦。

整体布局力求疏密有致，在小区中心营造花园，尽量降低建筑覆盖率，结合底层架空的手法，还地面以大量的绿地。采用"大花园"的规划原则，使水景、绿景向小区内部延伸，互补关联，力求水景利用最大化、水景住宅最多化和小区整体内外环境景观的均好性，使水韵、绿意无处不在。

空间形态设计从整体出发，立意在先，以人为本，强调人与自然的亲近，重塑传统的人文社区，注重小区内在环境的隐私性。每个组团均由住宅围合而成，形成了邻里型的私家花园。同时通过对入口的设计，对动静区的划分，形成了一个有私密性的公共活动空间，称之为小绿洲，这是一个只有参与围合的住宅居民才能进入的空间，形成私家花园。

根据景观对高层住宅设计的重要性，总体布局对景观特别注重，通过对各种景观系统分析，进而进行住宅空间布局级配，景观资源利用最大化与均好性并重的原则为本方案的一大特色。

建筑面积：145000m²
Total Floor Area: 145000m²
设计时间：2006年
Design Period: 2006

The project whose uptown area is 149614.2m² and which located in Zhabei district was divided into east and west parts by Pingxingguan Road. The area, far and construction height of the east part are 108300m², 2.5 and 100meters respectively.

The whole layout pursues proper density. The combination of debasing construction coverage ratio and the tact that built on stilts in bottom gave large virescence system to the ground. The "large garden" programming principle made the waterscape and virescence extend to district. Maximization of the waterscape, waterscape residences, and all good made the taste of water and green spread everywhere.

The design space design stressed the close of human and nature, remoulded the traditional cultural community, and emphasized the secret of the inner environment. The enclosure of the residences formed the neighbourhood private garden. A little oasis was formed by secret public space for activity, and it is a space of private admits only residents.

The whole layout stressed the sight. The principle of maximization the use of sight and all good is a significant feature of the project.

总平面图

阳江核电办公科研后勤基地规划
PLANNING FOR YANGJIANG NUCLEAR ENERGY ADMINISTRATION, RESEARCH, AND LOGISTICS BASE

基地位于阳江市区北郊共青湖水库以北、石塘山以南，东面为拟建体育北路、西面为拟建康泰路，基地依山傍水，朝向极佳，周边自然资源丰富。项目的主要设计特点如下：

1. "一轴、一中心、两翼街区"结构

"一轴"指对中部的车行主轴；"一中心"指办公科研中心区；"两翼街区"为办公科研区两侧后勤居住地块。

2. "显山露水"

基地整体空间形成"西高东低、北高南低"的空间形态，中部办公核心区规划为尺度巨大的开放空间，最大限度地把外部景观引入区内，成为山水交融相通的景观通廊，达到"显山露水"的效果。

3. "三园合一"

北面的石塘山规划为山林康体公园，南面共青湖规划为水景休憩乐园，与小区生态花园"三园合一"。利用地势和自然资源，保持低密度、高绿化率的原则，确保"公园化"的全面表现，实现"办公科研、生态居住、体育健康、游憩娱乐"的现代企业办公科研居住一体化乐园。

建筑面积：700000m²
Total Floor Area: 700000m²
设计时间：2006年
Design Period: 2006

The site is located in the north of Yangjiang, with excellent views and an optimal orientation. The main design specialities of the project are the following:

1. Planning structure is formulated as: "one axis, one centre, two limbs"
Where "One axis" is the main roadway connecting all major nodes within the project; "one centre" is the scientific research hub; and "two limbs" are the logistics plots on both sides of the administration and research centre.

2. Sightlines towards the mountain and water resource. The plot is elevated on its west and northern sides while it slopes down towards the east and north. The large open space adjacent to the central office area conveys the sights into the district, and forms the axis that optimizes sightlines towards the water and mountain range.

3. "Three gardens merge as a one".
It is formed by the combination of a fitness park, a waterscape fairyland, and an ecological garden in the district. Its low-density, large open space principle ensures a park-like effect throughout the project and realizes the integration of scientific research with ecology, fitness, well-being, rest and relaxation.

浙江湖州仁皇山新区城市设计
URBAN DESIGN OF HUZHOU RENHUANGSHAN NEW DISTRICT

该项目位居旧城与太湖间未建成区的山水环抱之中。新区的规划建设对城市北扩、对21世纪新湖州具有非同寻常的奠基与示范作用。国际竞标包含三个层次的设计内容：2.33km² 的城市设计；50hm² 的行政核心区设计；行政中心（6万m²）建筑设计和会展中心（3.5万m²）、科技文化中心（2.5万m²）概念设计。

设计从宏观入手，以深入调研为手段，以尊重生态环境、前瞻性思考及可持续发展为原则，力求达到自然景观与人文精神的统一、环境效益与社会效益的双赢，展现城市设计的新思维、新概念。

建筑面积：2330000m²
Total Floor Area：2330000m²
设计时间：2000 年
Design Period：2000

This project is situated between the old city and the Tai Lake and enclosed by mountain and water. Urban Planning for the new district acts as both foundation for urban renewal, and reference for city expansion for the new Huzhou city of 21st century. The international competitive tender calls for designs on three sectors: a) urban design for an area of 2.33km²; b) site layout for an administrative central district with an area of 50 hectares; c) architectural design for an administration centre (60000m²), an exhibition centre (35000m²), and the concept scheme for a technological centre (25000m²).

Investing careful analysis from macrostructure level down, the design strives to address ecological concerns and aims for sustainable development. It also tries to achieve multiple goals of unification between natural landscape and human spirit, environmental and social benefits, and expresses new concepts in urban design.

青岛李沧区下王埠村
LICANG XIAWANGBU VIALLAGE, QINGDAO

本项目位于李沧区的东北部，西北紧邻老虎山森林公园；东北背依崂山余脉卧龙山。现状用地呈现自然的村镇形态，土地开发利用程度较低，基础设施比较缺乏，可供开发利用的用地面积较大，改造可行性较强。308国道西侧地块地势稍有起伏，北面地块有一占地10000m² 的天然湖泊。

鉴于本项目的区位和总体定位，对如何体现青岛的门户形象，如何在规划中体现欧洲小镇风格，如何最大限度地利用青山绿水，就成为塑造小区品质个性的三个现实依据。设计中要求依据总体规划和基地环境条件，完善青岛李沧区虎山路片区的各项公共配套设施，提高整体环境质量水平，建设成生态、健康、符合中国市场的欧洲风情小镇。住宅设计上以奥地利的风情小镇——萨尔茨堡为概念。公共服务设施规划充分考虑不同配套设施的服务半径、步行距离保证的步行半径，合理配置不同功能的公建、配套设施。

设计以先进的居住区规划理念为指导，借鉴国内外成功的居住区项目的经验，创造生活便利，配套齐全，环境优美，个性突出，以21世纪的社会、经济、文化、环境和技术为设计依据，以人为本、生态环境为主，复兴城市街巷生活的生态型、人性化、现代化、国际化的居住社区。达到经济效益、社会效益和环境效益的统一。

建筑面积：362734m²
Total Floor Area：362734m²
设计时间：2005年
Design Period：2005

The site is located in the north eastern sector of Licang district, adjacent to the Tiger forest park to the northwest and the Wolong Mountain to the northeast. West of the site is an undulating national highway and to its north is a moderate sized natural lake. The existing site has a traditional village with relatively low development and poor infrastructure.

The planning commission requires significant improvement for the environment and an upgraded and efficient public facility devoted to human well being and responsive to ecological preservation. The project aims to create a residential community befitting international standards, taking inspiration from European community settings, particularly that of Salzburg, Austria.

厦门洪文居住区
HONGWENJU RESIDENCES, XIAMEN

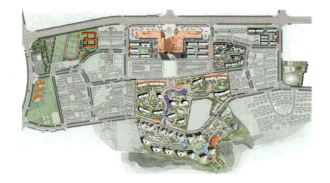

该项目位于厦门岛最高峰云顶岩北麓。西以厦门市东北部分区规划中的县黄路为界；北临城市东西主干道莲前路；东至紫薇花园，磨心山一带；南靠云顶岩山麓。

本项目沿莲前东路布置商业性公共设施用地，同时在各组团布置少量的便民店以及必备的公建配套。回迁安置房布置在商业中心两侧及莲前路东侧，小学及中学布置在沿县黄路一侧。

住宅立面设计以简洁明快为基本原则。在强调平面功能的同时，尽量避免多余的装饰，体现现代高新技术和信息社会的时代特征，寻求建筑造型艺术上的创新及与传统文化内涵及符号的巧妙结合。

塔楼部分用凸窗、转角窗、落地窗的设计手法，运用竖向和横向板带线条将阳台凸窗连成一个整体。主墙面为白色调，错落凹凸石墙面为浅灰色，利用线条与色彩取得与周边已建建筑的协调统一，形成整体的居住区印象。

建筑面积：718700m²
Total Floor Area: 718700m²
设计时间：2004 年
Design Period: 2004

This project is situated at the foot of the highest mountain in Xiamen, Yundingyan. The site is bounded by Xianhuang road on its west, Lianqian road on its north, Ziwei garden and Moxin Mountain on its east, as it leans against the foot of Yundingyan mountain in the south.

Public facility plots are distributed along Lianqian road, where each group is allocated with a commercial strip, and other necessary facilities. Storage houses are located on both sides of the commercial centre and at the east side of Lianqian road. A primary school and a middle school are located along Yanxian Rd.

Simplicity and minimalism are the basic principles for elevation design. Design is functional and excessive ornamentation is eschewed to accentuate its contemporary theme and a hi-tech feel. It also strives for a working coexistence between modern design, innovative construction and traditional elements for a more efficient integration into the community on which the project site is located.

Panels of bay windows, corner windows, and French windows are used in the towers to combine fenestrations and balconies into an integrated whole. White is the main colour scheme which is accessorized with occasional greys, mouldings and contours to create refreshing new residential area.

厦航同安 T2006G01 地块项目
XIAHANG TONG AN T2006G01 PLOT PROJECT

同安T2006G01地块位于厦门同安，东侧为T2006G02拍卖地块；南侧为碧溪路和东西溪；东西侧7m宽规划路；北侧为已形成24m宽凤山路。建设用地南北宽约190m；东西长约330m。平面近似长方形。用地性质住宅＋商业；地形平坦，视野开阔。

在满足规划条件及业主具体要求前提下，本着合理利用土地条件的原则，将低层住宅布置在临溪的南侧，小高层布置其后，而将高层主要布置在北侧，从溪边开始逐步增加建筑高度，充分利用景观和朝向，同时尽量增加室外活动空间。底商与部分两层集中商业相结合，以商铺的形式为主，沿主要人流方向布置在凤山路一侧及东西两条规划路。保留地块入口处的古榕树，作为人文、历史的延续。

整体建筑风格采用现代风格的设计手法，追求自然简洁，反映居住与自然的良好关系。主墙面采用浅色，错落和凹凸的墙面穿插色彩变化，对比中保持相同，灵活运用空中庭园、凸窗、转角窗的交错使用，为简洁立面带来了丰富元素。

建筑面积：142866m²
Total Floor Area：142866m²
设计时间：2006 年
Design Period：2006

The project is proposed as a residential-commercial district in Tong'an, Xiamen. The site area is ellipsoid with a relatively flat contour and wide vistas all around.

In order to satisfy programming requirements and using practical site layout as the backbone for planning, low-rise structures are located in the south with building heights gradually rising as structures are dispersed towards the north. This tact improves sight corridors and enlarges spaces for outdoor activities.

The design adopts a contemporary theme of simplicity and minimalism to highlight a harmonious coexistence between man and nature. Hanging courtyards, bay windows, and a variable colour scheme provide stimulating features for the simple elevation.

山东农业银行大楼
SHANDONG AGRICULTURE BANK BUILDING

建筑面积：15600m²
Total Floor Area：15600m²
设计时间：1996 年
Design Period：1996
竣工时间：1999 年
Completion Period：1999

绵阳中级人民法院
INTERMEDIATE PEOPLE'S COURT, MIANYANG

建筑面积：24589m²
Total Floor Area：24589m²
设计时间：1999 年
Design Period：1999

深圳金田大厦
JINTIAN BUILDING, SHENZHEN

建筑面积：23400m²
Total Floor Area：23400m²
设计时间：1991 – 1992 年
Design Period：1991 – 1992
竣工时间：1994 年
Completion Period：1994

深圳龙岗区政府大楼
LONGGANG DISTRICT ADMINISTRATIVE OFFICE BUILDING, SHENZHEN

建筑面积：54596m²
Total Floor Area：54596m²
设计时间：1993 年
Design Period：1993
竣工时间：1995 年
Completion Period：1995

苍南行政中心
CANGNAN ADMINISTRATIVE CENTER

建筑面积：57262m²
Total Floor Area：57262m²
设计时间：2001 – 2004 年
Design Period：2001 – 2004

山东省国税局综合业务楼
SHANDONG STATE INTEGRATED TAX BUREAU BUILDING

建筑面积：29066.14m²
Total Floor Area：29066.14m²
设计时间：2001 年
Design Period：2001

深圳盐田国际行政大楼
YANTIAN INTERNATIONAL ADMINISTRATION BUILDING, SHENZHEN

建筑面积：23243.5m²
Total Floor Area：23243.5m²
设计时间：2001 年
Design Period：2001

南宁行政中心
NANNING ADMINISTRATIVE CENTER

建筑面积：50000m²
Total Floor Area：50000m²
设计时间：2002 年
Design Period：2002

深圳龙岗规划展览综合大楼
LONGGANG DISTRICT EXHIBITION BUILDING, SHENZHEN

建筑面积：26126m²
Total Floor Area：26126m²
设计时间：2003 年
Design Period：2003

深圳市地方税务局第六检查分局大楼
TAX BUREAU BUILDING, SIXTH BRANCH, SHENZHEN

建筑面积：23657m²
Total Floor Area：23657m²
设计时间：2003 – 2004 年
Design Period：2003 – 2004

澳门中联办行政办公大楼
MACAO ZHONGLIANBAN ADMINISTRATIVE OFFICE BLOCK

建筑面积：23000m²
Total Floor Area：23000m²
设计时间：2005年
Design Period：2005

深圳天健大厦
TIANJIAN BUILDING, SHENZHEN

建筑面积：36389.29m²
Total Floor Area：36389.29m²
设计时间：2004年
Design Period：2004

深圳中学文体楼
SHENZHEN MIDDLE SCHOOL CULTURAL & PHYSICAL TRAINING BUILDING

建筑面积：8000m²
Total Floor Area：8000m²
设计时间：2000年
Design Period：2000
竣工时间：2002年
Completion Period：2002

深圳华侨城九年一贯制学校
SHENZHEN OVERSEAS CHINESE SCHOOL

建筑面积：24000m²
Total Floor Area：24000m²
设计时间：2001年
Design Period：2001

北大附中广州分校光大校区
MIDDLE SCHOOL AFFILIATED TO BEIJING UNIVERSITY GUANGZHOU BRANCH EVERBRIGHT CAMPUS

总建筑面积：32800m²
Total Floor Area：32800m²
设计时间：2001年
Design Period：2001

哈尔滨工业大学深大校区
HAERBIN INSTITUTE OF TECHNOLOGY SHENZHEN UNIVERSITY CAMPUS

建筑面积：92940.45m²
Total Floor Area：92940.45m²
设计时间：2002 年
Design Period：2002

东莞理工学院松山湖校区
DONGGUAN UNIVERSITY OF TECHNOLOGY SONGSHANHU CAMPUS

建筑面积：15500m²
Total Floor Area：15500m²
设计时间：2002 – 2003 年
Design Period：2002 – 2003

绵阳中学小岛实验学校
MIANYANG MIDDLE SCHOOL

建筑面积：67070m²
Total Floor Area：67070m²
设计时间：2003 年
Design Period：2003

深大附中文体楼
SHENZHEN MIDDLE SCHOOL CULTURE & PHYSICAL TRAINING BUILDING

建筑面积：6014m²
Total Floor Area：6014m²
设计时间：2005 年
Design Period：2005

深圳景轩酒店
SHENZHEN JINGXUAN HOTEL

建筑面积：28338m²
Total Floor Area：28338m²
设计时间：1993 年
Design Period：1993
竣工时间：1998 年
Completion Period：1998

惠州西湖山庄
HUIZHOU WEST LAKE MOUNTAIN VALLA

建筑面积：36100m²
Total Floor Area：36100 m²
设计时间：2005 年
Design Period：2005

香港沙田宝福灵堂
HONGKONG SHATSIM POFOOK MEMORIAL HALL

建筑面积：2200m²
Total Floor Area：2200m²
设计时间：1986-1987 年
Design Period：1986 – 1987
竣工时间：1989 年
Completion Period：1989

东莞中心汽车客运站
DONGGUAN CENTRAL BUS STATION

建筑面积：21784m²
Total Floor Area：21784m²
设计时间：2000 年
Design Period：2000

江苏省电视台
JIANGSU PROVINCIAL TV STATION

建筑面积：88496m²
Total Floor Area：88496m²
设计时间：1998 年
Design Period：1998

深圳火车站
SHENZHEN RAILWAY STATION

建筑面积：93790m²
Total Floor Area：93790m²
设计时间：1989 年
Design Period：1989
竣工时间：1991 年
Completion Period：1991

广东南海影剧院
GUANGDONG NANHAI THEATER

建筑面积：8574m²
Total Floor Area：8574m²
设计时间：1998 年
Design Period：1998
竣工时间：1999 年
Completion Period：1999

温州大剧院
WENZHOU LARGE SHOWPLACE

建筑面积：30600m²
Total Floor Area：30600m²
设计时间：2000 年
Design Period：2000

深圳老干部活动中心
SHENZHEN ELDERS ACTIVITY CENTER

建筑面积：26257m²
Total Floor Area：26257m²
设计时间：2005 年
Design Period：2005

南海西樵科技信息会展中心
NANHAI XIQIAO CONVENTION & EXHIBITION CENTER OF SCIENCE & TECHNOLOGY

建筑面积：22000m²
Total Floor Area：22000m²
设计时间：1999 年
Design Period：1999

南京国际展览中心
NANJING INTERNATIONAL EXHIBITION CENTER

建筑面积：138100m²
Total Floor Area：138100m²
设计时间：1998 年
Design Period：1998

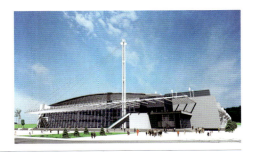

深圳鸿展商城
SHENZHEN HONGZHAN MALL

建筑面积：14777m²
Total Floor Area：14777m²
设计时间：1999－2000 年
Design Period：1999－2000
竣工时间：2003 年
Completion Period：2003

深圳金世界商业中心
GOLD WORLD COMMERCIAL CENTER

建筑面积：14898m²
Total Floor Area：14898 m²
设计时间：1999 年
Design Period：1999

深圳东门风貌街
SHENZHEN DONGMEN FENGMAO STREET

建筑面积：16000m²
Total Floor Area：16000m²
设计时间：1998－1999 年
Design Period：1998－1999
竣工时间：2000 年
Completion Period：2000

成都科技一条街
CHENGDU SCIENCE & TECHNOLOGY STREET

规划用地：15ha
Total Floor Area：15ha
设计时间：2003－2004 年
Design Period：2003－2004

深圳国贸商住大厦
INTERNATIONAL TRADE COMPOSITE BUILDING, SHENZHEN

建筑面积：24223m²
Total Floor Area：24223m²
设计时间：1987－1988 年
Design Period：1987－1988
竣工时间：1990 年
Completion Period：1990

深圳国贸新商住大厦
NEW INTERNATIONAL TRADE COMPOSITE BUILDING, SHENZHEN

建筑面积：52600m²
Total Floor Area：52600m²
设计时间：1990 年
Design Period：1990
竣工时间：1995 年
Completion Period：1995

湛江市移动通信综合楼
ZHANJIANG MOBILE COMMUNICATIONS INTEGRATED BUILDING

建筑面积：13304m²
Total Floor Area：13304m²
设计时间：1999-2000 年
Design Period：1999 – 2000
竣工时间：2004 年
Completion Period：2004

绵阳临园大厦
MIANYANG LINYUAN BUILDING

建筑面积：37749m²
Total Floor Area：37749m²
设计时间：1999 年
Design Period：1999
竣工时间：2000 年
Completion Period：2000

绵阳樊华大厦
MIANYANG XIANGHUA BUILDING

建筑面积：133000m²
Total Floor Area：133000m²
设计时间：1997 年
Design Period：1997

广西玉林凯旋广场
GUANGXI YULIN TRIUMPH PLAZA

总建筑面积：66525.9m²
Total Floor Area：66525.9m²
设计时间：2006 年
Design Period：2006

深圳嘉汇新城
SHENZHEN JIAHUI NEW BUILDING

建筑面积：139752m²
Total Floor Area：139752m²
设计时间：1995 年
Design Period：1995
竣工时间：2000 年
Completion Period：2000

合肥招商大厦
HEFEI MERCHANTS BUILDING

建筑面积：100000m²
Total Floor Area：100000m²
设计时间：2005 年
Design Period：2005

日本奈良中国文化村
NARA CHINESE CULTURAL VILLAGE, JAPAN

建筑面积：30000m²
Total Floor Area：30000m²
设计时间：1986 – 1994 年
Design Period：1986 – 1994

加拿大枫华苑酒店
SINOMONDE GARDEN HOTEL

建筑面积：30000m²
Total Floor Area：30000m²
设计时间：1988 – 1989 年
Design Period：1988 – 1989
竣工时间：1991 年
Completion Period：1991

山东蓬莱利群酒店
SHANDONG PENGLAILIQUN HOTEL

建筑面积：59798m²
Total Floor Area：59798m²
设计时间：2005 年
Design Period：2005

南充罗曼威森国际商务酒店
NANCHONG ROMANVISEN
INTERNATIONAL COMMERCIAL HOTEL

建筑面积：22000m²
Total Floor Area: 22000m²
设计时间：2006 年
Design period：2006

宁波康城阳光酒店
NINGBO CANNES SUNSHINE HOTEL

建筑面积：55028m²
Total Floor Area：55028m²
设计时间：2005 年
Design Period：2005

湖南省游泳跳水中心
SWIMMING & DIVING CENTER, HUNAN

建筑面积：18360m²
Total Floor Area：18360m²
设计时间：2001 年
Design Period：2001

贵阳白云区行政中心
BAIYUN DISTRICT ADMINISTRATIVE
CENTER, GUIZHOU

建筑面积：39678.06m²
Total Floor Area：39678.06m²
设计时间：2001 – 2003 年
Design Period：2001 – 2003

南海移动通信局办公楼
MOBILE COMMUNICATIONS BUILDING, NANHAI

建筑面积：12688m²
Total Floor Area：12688m²
设计时间 1999 – 2000 年
Design period：1999 – 2000
竣工时间：2004 年
Completion Period：2004

武汉东湖宾馆别墅
WUHAN EAST LAKE HOTEL VILLA

建筑面积：4646m²
Total Floor Area：4646m²
设计时间：1999 年
Design Period：1999

无锡锦园宾馆改扩建工程
RENOVATION AND EXTENSION PROJECT OF WUXI JINYUAN HOTEL

建筑面积：85700m²
Total Floor Area：85700m²
设计时间：2000 年
Design Period：2000

北京未来家园
BEIJING FUTURE HOME

建筑面积：660400m²
Total Floor Area：660400m²
设计时间：2001 年
Design Period：2001

天津老城厢地块规划
TIANJIN ANCIENT CITY PLOT LAYOUT

建筑面积：213358.9m²
Total Floor Area：213358.9m²
设计时间：2004 年
Design Period：2004

济南千佛山南麓项目
JINAN QIANFO MOUNTAIN PROJECT

建筑面积：267000m²
Total Floor Area：267000m²
设计时间：2004 年
Design Period：2004

青岛四方美茵茨花园
QINGDAO SQUARE BEAUTIFUL GARDEN

建筑面积：207150m²
Total Floor Area：207150m²
设计时间：2003 年
Design Period：2003

无锡 xdg-2004 地块规划
WUXI XDF-2004 PLOT PLANNING

建筑面积：426513m²
Total Floor Area：426513m²
设计时间：2006 年
Design Period：2006

温州杨府山居住区规划
WENZHOU YANGFU MOUNTAIN RESIDENCES LAYOUT

建筑面积：661730m²
Total Floor Area：661730m²
设计时间：2000 年
Design Period：2000

广州万科四季花城
GUANGZHOU VANKE WONDER LAND

建筑面积：200000m²
Total Floor Area：200000m²
设计时间：2003 年
Design Period：2003

中山鄂尔多斯
ZHONGSHAN EERDUOSI

建筑面积：343423m²
Total Floor Area：343423m²
设计时间：2003 年
Design Period：2003

南宁·塞纳维拉
NANNING SAINA VILLA

建筑面积：395019m²
Total Floor Area：395019m²
设计时间：2003 – 2004 年
Design Period：2003 – 2004

长沙天健芙蓉中心路规划
CHANGSHA TIANJIAN LOTUS RD PROJECT

建筑面积：769281.39m²
Total Floor Area：769281.39m²
设计时间：2004 年
Design Period：2004

长沙蔚蓝海岸
CHANGSHA BLUE COAST

建筑面积：293718.5m²
Total Floor Area：293718.5m²
设计时间：2004 年
Design Period：2004

南昌联泰香郁滨江
LIANTAI XIANGYU SEASIDE, NANCHANG

建筑面积：1500000m²
Total Floor area：1500000m²
设计时间：2006 年
Design Period：2006

成都国际社区
CHENGDU INTERNATIONAL COMMUNITY

建筑面积：1143131m²
Total Floor Area：1143131m²
设计时间：2004 年
Design Period：2004

丹阳怡康家园
DANYANG YIKANG GARDEN

建筑面积：130311.6m²
Total Floor Area：130311.6m²
设计时间：2005 年
Design Period：2005

深圳朗矩创业 SOHO 城
LANGJU SOHO CITY, SHENZHEN

建筑面积：111381m²
Total Floor Area：111381m²
设计时间：2003 年
Design Period：2003

长春金色家园
CHANGCHUN GOLDEN HOMESTEAD

建筑面积：61105m²
Total Floor Area：61105m²
设计时间：2000 年
Design Period：2000

北京柏联别墅
BEIJING BOLIAN VILLA

建筑面积：87928.8m²
Total Floor Area：87928.8m²
设计时间：2002 年
Design Period：2002

天津万春花园
TIANJIN WANCHUN GARDEN

建筑面积：140000m²
Total Floor Area：140000m²
设计时间：1998－1999 年
Design Period：1998－1999
竣工时间：2002 年
Completion Period：2002

济南港澳花园
JINAN GANGAO GARDEN

建筑面积：55000m²
Total Floor Area：55000m²
设计时间：2005 年
Design Period：2005

盐城黄海明珠苑
YANCHENG YELLOW SEA BRIGHT PEARL GARDEN

建筑面积：125274m²
Total Floor Area：125274m²
设计时间：2005
Design Period：2005

泉州千亿山庄别墅区及商业街
QUANZHOU QIANYI VILLAS DISTRICT AND COMMERCIAL STREET

建筑面积：118800m²
Total Floor Area：118800m²
设计时间：2000 年
Design Period：2000

深圳美加广场
SHENZHEN MEIJIA PIAZZA

建筑面积：59063m²
Total Floor Area：59063m²
设计时间：1996 年－1997 年
Design Period：1996－1997
竣工时间：1999 年
Completion Period：1999

南海怡翠花园
NANHAI YICUI GARDEN

建筑面积：480000m²
Total Floor Area：480000m²
设计时间：1998－1999 年
Completion Period：1998－1999

深圳万事达名苑（二期）
SHENZHEN WANSHIDA FAMOUS GARDEN

建筑面积：41000m²
Total Floor Area：41000m²
设计时间：1999 年
Design Period：1999

深圳华侨城中心花园
SHENZHEN HUAQIAO CITY CENTRAL GARDEN

建筑面积：41023m²
Total Floor Area：41023m²
设计时间：2000 年
Design Period：2000

深圳紫薇苑住宅小区
SHENZHEN ZIWEI RESIDENTIAL DISTRICT

建筑面积：46398m²
Total Floor Area：46398m²
设计时间：1999 年
Design Period：1999

深圳大鹏海滨度假别墅
SHENZHEN DAPENG SEASIDE HOLIDAY VILLA

建筑面积：1395.1m²
Total Floor Area：1395.1m²
设计时间：2001 年
Design Period：2001
竣工时间：2003 年
Completion Period：2003

深圳绿景新美域
SHENZHEN GREEN VIEW GARDEN

建筑面积：165722m²
Total Floor Area：165722m²
设计时间：2002 - 2003 年
Design Period：2002 - 2003

深圳中信·海阔天空
SHENZEHN HAIKUOTIANKONG GARDEN

建筑面积：97726.62m²
Total Floor Area：97726.62m²
设计时间：2002年 — 2003年
Design Period：2002 — 2003
竣工时间：2004年
Completion Period：2004

深圳华茂欣园
SHENZHEN HUAMAOXIN GARDEN

建筑面积：53622.4m²
Total Floor Area：53622.4m²
设计时间：2002 — 2003年
Design Period：2002 — 2003
竣工时间：2004年
Completion Period：2004

深圳友邻国际公寓
SHENZHEN SEAVIEW GARDEN

建筑面积：56139.4m²
Total Floor Area：56139.4m²
设计时间：2003 — 2004年
Design Period：2003 — 2004

深圳东方华都
SHENZHEN JINGJI CITY APARTMENT

建筑面积：50000m²
Total Floor Area：50000m²
设计时间：2003年
Design Period：2003

深圳布吉诚信华庭
SHENZHEN BUJI CHENGXINHUATING

建筑面积：31733.7m²
Total Floor Area：31733.7m²
设计时间：2004年
Design Period：2004

东莞中惠丽阳时代
DONGGUAN ZHONGHUI LIYANG TIMES

建筑面积：100000m²
Total Floor Area：100000m²
设计时间：2003 – 2004 年
Design Period：2003 – 2004
竣工时间：2005 年
Completion Period：2005

南海金沙湾
NANHAI GOLDEN BAY

建筑面积：192873m²
Total Floor Area：192873m²
设计时间：2005 年
Design Period：2005

昆明万辉西山新城
KUNMING WANHUI WEST MOUNTAIN NEW CITY

建筑面积：3000000m²
Total Floor Area：3000000m²
设计时间：2006 年
Design Period：2006

贵阳"在水一方"
GUIYANG "IN THE WATER"

建筑面积：121708m²
Total Floor Area：121708m²
设计时间：2001 年
Design Period：2001

湖南常德紫金城
CHANGDE ZIJIN TOWN

建筑面积：150000m²
Total Floor Area：150000m²
设计时间：2004 年
Design Period：2004

长沙湘麓国际
CHANGSHA XIANGLU INTERNATIONAL

建筑面积：201000m²
Total Floor Area：201000m²
设计时间：2005 – 2006
Design Period：2005 – 2006

武汉樱花大厦
WUHAN CHERRY BLOSSOM MANSION

建筑面积：28000m²
Total Floor Area：28000m²
设计时间：2001 – 2002 年
Design Period：2001 – 2002

深圳金湖山庄
SHENZHEN GOLD LAKE MOUNTAIN VILLA

建筑面积：32100m²
Total Floor Area：32100m²
设计时间：1991 年
Design Period：1991
竣工时间：1992 年
Completion Period：1992

河源长城·世纪华府
GREAT WALL SHIJI HUAFU, HEYUAN

建筑面积：141430m²
Total Floor Area：141430m²
设计时间：2005
Design Period：2005

重庆骏逸天下
CHONGQING JUNYI TIANXIA

建筑面积：202600m²
Total Floor Area：202600m²
设计时间：2002 – 2003 年
Design Period：2002 – 2003
竣工时间：2004 年
Completion Period：2004

重庆朵力现代城
CHONGQING DUOLI MODERN CITY

建筑面积：337970m²
Total Floor Area：337970m²
设计时间：2004 年
Design Period：2004

成都天府长城三期
STAGE III OF CHENGDU TIANFU GREAT WALL

建筑面积：290000m²
Total Floor Area：290000m²
设计时间：2006 年
Design Period：2006

四川德阳文庙广场
SICHUAN DEYANG WEN TEMPLE PLAZA

建筑面积：82236m²
Total Floor Area：82236m²
设计时间：2002 – 2003 年
Design Period：2002 – 2003
竣工时间：2004 年
Completion Period：2004

莆田·第一城
PUTIAN · SPLENDID CITY

建筑面积：350000m²
Total Floor Area：350000m²
设计时间：2005 年
Design Period：2005

南京苏源颐和美地（西园）
SUYUAN YIHE MEIDI (WEST GARDEN), NANJING

建筑面积：210000m²
Total Floor Area：210000m²
设计时间：2001 年
Design Period：2001
竣工时间：2002 年
Completion Period：2002

南充地中海蓝酒店公寓
NANCHONG MEDITERRANEAN BLUE HOTEL APARTMENT

建筑面积：31000m²
Total Floor Area：31000m²
设计时间：2005
Design Period：2005

江西丰城金马·城市之都
FENGCHENG JINMA CAPITAL OF THE CITY, JIANGXI

建筑面积：227045m²
Total Floor Area: 227045m²
设计时间：2005
Design Period：2005

合肥学府·星座公寓
HEFEI ACADEMY CONSTELLATION FLAT

建筑面积：31933m²
Total Floor Area：31933m²
设计时间：2005 年
Design Period：2005

重庆华宇名都
CHONGQING HUAYU MINGDU

建筑面积：316412m²
Total Floor Area：316412m²
设计时间：2001 年
Design Period：2001

青岛锦绣海岸
QINGDAO JINXIU SEACOAST

建筑面积：120000m²
Total Floor Area：120000m²
设计时间：2002－2004 年
Design Period：2002－2004

青岛浮山后静湖琅园
FUSHANHOU JINGHULANG GARDEN, QINGDAO

建筑面积：120000m²
Total Floor Area：120000m²
设计时间：2004 年
Design Period：2004

青岛十梅庵片区规划
SHIMEIAN DISTRICT PLANNING, QINGDAO

规划面积：540ha
Total Floor Area：540ha
设计时间：2004 年
Design Period：2004

图书在版编目(CIP)数据

华艺设计 1986—2006/华艺设计顾问有限公司编.
北京：中国建筑工业出版社，2006
ISBN 7-112-08686-8

Ⅰ.华… Ⅱ.华… Ⅲ.建筑设计－作品集－中国
－1986—2006　Ⅳ.TU206

中国版本图书馆CIP数据核字(2006)第123632号

责任编辑：王莉慧　曹　扬
版式设计：傅金红
责任校对：邵鸣军　王雪竹

华艺设计
1986—2006
华艺设计顾问有限公司　编

*

中国建筑工业出版社出版、发行(北京西郊百万庄)
新华书店经销
北京广厦京港图文有限公司设计制作

*

开本：889×1194毫米　1/12　印张：22$\frac{2}{3}$　字数：690千字
2006年10月第一版　2006年10月第一次印刷
印数：1—3500册　定价：268.00元
ISBN 7-112-08686-8
　　　(15350)

版权所有　翻印必究
如有印装质量问题,可寄本社退换
(邮政编码　100037)
本社网址：http://www.cabp.com.cn
网上书店：http://www.china-building.com.cn